THE GREAT SCIENTISTS

John Farndon

and

Alex Woolf · Anne Rooney · Liz Gogerly

Arcturus Publishing Limited
26/27 Bickels Yard
151–153 Bermondsey Street
London SE1 3HA

Published in association with
foulsham
W. Foulsham & Co. Ltd,
The Publishing House, Bennetts Close, Cippenham,
Slough, Berkshire SL1 5AP, England

ISBN 0-572-03148-3

This edition printed in 2005

British Library Cataloguing-in-Publication Data: a catalogue record for this book is available from the British Library

Printed in China

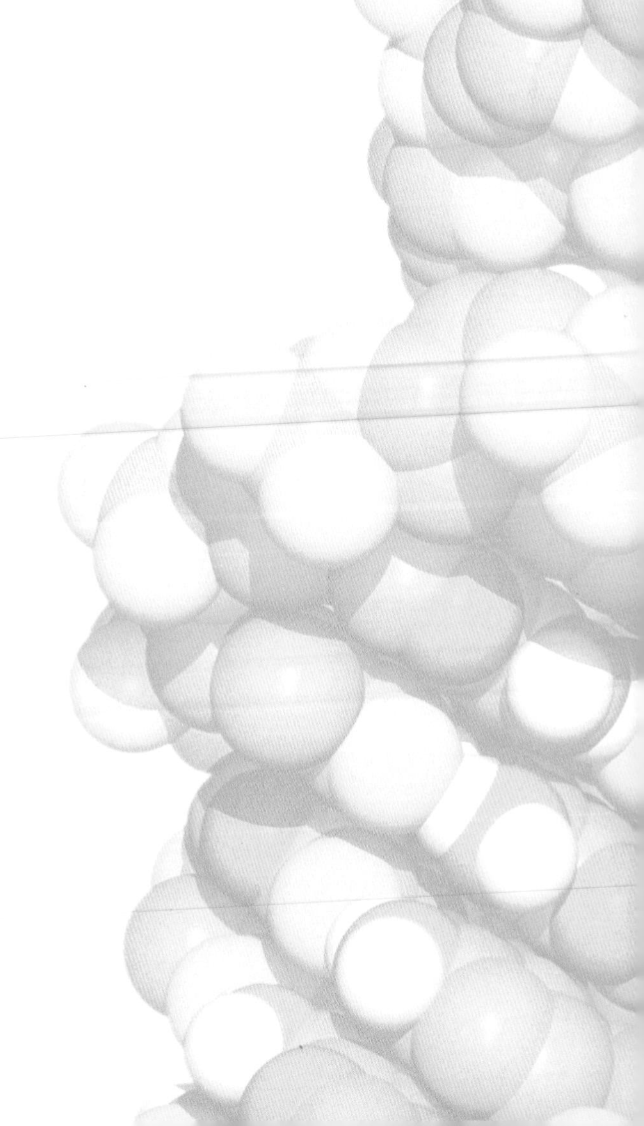

Contents

Introduction

Science has been around a long time, at least since the days of the ancient Greeks. Indeed, if one consults the archaeological record, it seems clear that the Babylonian and Sumerian civilisations had rather more than a rudimentary grasp of medicine, astronomy and applied mathematics, not to mention engineering.

From around the sixth century BC, however, we see signs in the ancient Greek world of what could perhaps be called the first scientific revolution. No longer satisfied with the gods as the ultimate answer to why the world is the way it is, Greek thinkers began to search for an underlying principle which would form the basis of a more satisfactory explanation. The great Thales of Miletus proposed that this prime substance was water; Anaxagoras believed it to be air; Xenophanes proposed the rather less glamorous option of mud. Democritus, astonishingly, proposed the first atomic theory – the word atom comes from the Greek atomon, literally translatable as 'indivisible' (just how inappropriate the word was would not be recognized until demonstrated in spectacular fashion by Ernest Rutherford in the twentieth century). What we read in the works of these pre-Socratic philosophers is not recognizable science, but we can see the glimmerings of the scientific method in the rejection of 'truth by authority', and the search for causes and principles based on observation and reason; truth as the province of thinkers, rather than of priests.

With the advent of Euclid and Archimedes, whose monumental works on geometry and trigonometry – among others – are still required reading on mathematics courses today, we find ourselves on the terra cognita of recognizable science.

The progress of science from then to the present day has not always been straightforward, however. Science has often come into conflict with organized religion, and on these occasions, the practitioners of science generally seem to come off worst, not least because the practitioners of religion seem all too eager to resort to threats, intimidation and even assassination in order to avoid hearing opinions they do not agree with. At certain periods in history, established churches have been eager to detect the odour of heresy in scientific accounts of the cosmos, the origin and structure of the earth, and especially the scientific account of the origins of humanity.

However battered and bruised it may at times have been, at the end of the twentieth century science has emerged the victor, as the key intellectual discipline for the twenty-first century and beyond. The reason for this success can be stated in two words: Science works. The whole laborious trial and error of the scientific process, the testing, revising and discarding of hypotheses; the diligent construction of theories which fit the known facts, and the modification or abandonment of these as and when new facts emerge; in short, the scientific method, produces results, results which are testable, verifiable, falsifiable, and from which predictions can be made. Astrology, cheiromancy, divination, oneiromancy, parapsychology, telepathy, UFO-logy, creation-'science', forecasting the weather from the entrails of sacrificial animals produce no such results. As has been remarked elsewhere, it is no accident that those societies which have actively embraced the scientific method have flourished, while those societies which have preferred instead to rely on superstition, witchcraft and religion have failed.

The men and women whose biographies make up this volume have all made outstanding contributions to their own fields of scientific endeavour, have all shed light onto more or less baffling phenomena, and have all contributed to the ever-expanding pool of human knowledge.

This book makes no claim, however, to be a comprehensive list: there are many more scientists who could have been included; by the same token, this book could run to many hundreds of pages and still be incomplete. Still less does it aim to spark a discussion on the nature of 'greatness' – such discussion can safely be left to others. What it does aim to do is to provide an insight into some of history's most influential scientific discoveries and discoverers, and to encourage the reader to take their interest further. Whether or not it succeeds in this aim, then, must be up to each indiviual reader to decide for themselves.

Euclid

Building on the work of early Greek philosophers such as Thales of Miletus and Anaximander, Euclid showed that events in the world could be understood by the application of reason, rather than by appeal to the gods.

c. 300BC

IT IS SAID THAT EUCLID'S GREAT BOOK THE *ELEMENTS* is the most widely translated, published and studied mathematical book in the western world. It is without doubt one of the greatest and most influential books of all time.

The *Elements* is basically about geometry, the mathematics of shape. It is such a thorough study that it remains the basic framework for geometry today, thousands of years after it was written. Mathematicians still refer to the geometry of flat surfaces – lines, points, shapes, and solids – as Euclidean geometry. In the *Elements* are summarized most of the basic rules of geometry, about triangles, squares, circles, parallel lines and so on that children learn at school today.

Euclid's great book also marked the birth of a whole new way of thinking, in which the way to truth can be found by logic, deductive reasoning, evidence and proof and not simply by leaps of intuition and faith. Now mankind no longer needed to regard the workings of the world as controlled by the whim of the gods, but as following natural rules that could gradually be discovered by using Euclid's methods.

However, this achievement was not Euclid's alone. He built on centuries of intellectual effort by Greek thinkers, dating back to the almost legendary Thales of Miletus in the seventh century BC. Yet Euclid's work encapsulated this approach to thinking in such a thorough and foolproof way that its lasting influence was guaranteed. Benedict de Spinoza, Immanuel Kant and Abraham Lincoln are among the countless people through history to have been inspired by his way of thinking.

Euclid the man

Very little is known about Euclid himself. It seems likely that he lived around 300BC in Alexandria, the great Egyptian city then newly founded by Alexander the Great on the shores of the Mediterranean. The first Greek ruler of Egypt, Ptolemy Soter (c. 367–283BC), created the Museum and Library in Alexandria, which became the most remarkable intellectual and educational institution in the ancient world, and Euclid was probably the leading mathematics teacher there. He may have been a student there under Plato, and Archimedes arrived there not long after Euclid died.

There are a few hints about his character from anecdote. It seems Euclid was a gentle and encouraging teacher. According to one source, he was 'most fair and well disposed towards all who were able in any measure to advance mathematics, careful in no way to give offence, and although an exact scholar never vaunting himself'. Another source tells about what happened when a student, frustrated by the effort of learning geometry, asked what he would get from studying it. In reply, apparently, Euclid called a servant, gave him some money and said, 'Give him these coins since he must make a gain out of what he learns.'

Another story tells how one day King Ptolemy asked if he had to read the whole of the *Elements* in order to learn about geometry, and Euclid diplomatically replied, 'There is no royal road to geometry'.

But this is pretty much all that is known. In fact, most of this comes from the writings of the Greek philosopher Proclus, who lived nearly 800 years later.

So little is known about Euclid that some scholars have suggested that the *Elements* could be the work of a team of scholars working under his guidance – and others even that 'Euclid' is simply the name a group of Alexandrian mathematicians gave themselves. But whatever the truth, there is no doubting the importance of the *Elements* and the other less-known works of Euclid.

Euclid and geometry

Euclid's great achievement was to combine the geometrical theorems of his day into a coherent framework of basic theory and proofs, which is the basis of all science to this day.

> Euclid's achievement was to create a coherent framework of basic theory and proofs, which forms the basis of all science to this day.

Geometry was already quite well developed by Euclid's time. Geometry is the mathematics of shape, and it probably began thousands of years earlier, arising from people's need to work out the area of their land. It was developed to a sophisticated level by the ancient Egyptians, who used it in the building of their pyramids. They called geometry 'earth measurement' and the Greeks adopted the term – the word 'geometry' is simply Greek for earth measurement. In 1858, Scottish historian Alexander Rhind found a papyrus scroll ('papyrus' is the name of the reed from which the Egyptians made their paper) written by an Egyptian scribe called Ahmes around 1650BC. The Rhind papyrus, and another papyrus now in Moscow (and so called the Moscow papyrus), showed that the ancient Egyptians knew a great deal about the geometry of triangles. For instance, they knew how to work out the height of things from the length of their shadow on the ground.

In fact, the Egyptians probably had a pretty good working knowledge of most of the geometric techniques described in the *Elements*. What Euclid and the ancient Greeks did was develop these practical techniques into a purely theoretical system, taking what might now be called 'applied mathematics' and creating what we would call 'pure mathematics'.

The Greeks searched for general abstract truths very much for their own sake, but what they discovered made their work far more important than an interesting intellectual pastime. Their method was such a powerful tool that the general truths it produced could be applied to every situation. What was true about triangles in one situation was true of them in another that was completely different. For instance, Thales of Miletus stunned the ancient Egyptians when he travelled there by showing how the method of similar triangles could be used to measure both the height of the pyramids and the distance of a ship at sea.

Euclid's windmill proof

Perhaps the most striking example of the power of Euclid's approach was his 'windmill' proof of Pythagoras's theory of right-angled triangles, so called because the diagrams looked like windmills. Indeed, it was so striking that in 1821 a German physicist suggested it could be the perfect demonstration of human intelligence to beings from other worlds. All we had to do to impress the Martians, it was claimed, was to plough and plant large fields in the shape of the windmill diagram, or dig huge canals in Siberia in the same shape, fill them with oil and set fire to them. Of course, no one has yet put this plan into action.

Both the ancient Egyptians and the Babylonians were completely familiar with the notion that the sides of a right-angle triangle were always in exactly the same proportion. They knew the length of each side was always in the same proportion to the 'squares' of the length of each of the other two sides – that is, the length multiplied by itself. Essentially, they knew what we now call the 'Pythagoras theorem' long before Pythagoras. This is the idea that adding together the squares of the two sides either side of the right angle gives the square on the third side, called the hypotenuse. What Pythagoras did in the sixth century BC was prove that this was so, but his proof was quite cumbersome. Euclid's windmill proof was simple and elegant:

1. Draw squares on the sides of the right ΔABC.
2. BCH and ACK are straight lines because <ACB = 90°
3. <EAB = <CAI = 90°, by construction.
4. <BAI = <BAC + <CAI = <BAC + <EAB = <EAC, by 3.
5. AC = AI and AB = AE, by construction.
6. Therefore, ΔBAI ΔEAC, as highlighted in part (a) of the figure.
7. Draw CF parallel to BD.
8. Rectangle AGFE = 2ΔACE. This remarkable result derives from two preliminary theorems: (a) the areas of all triangles on the same base, whose third vertex lies anywhere on an indefinitely extended line parallel to the base, are equal; and (b) the area of a triangle is half that of any parallelogram (including any rectangle) with the same base and height.
9. Square AIHC = 2ΔBAI, by the same parallelogram theorem as in step 8.
10. Therefore, rectangle AGFE = square AIHC, by steps 6, 8, and 9.
11. <DBC= <ABJ, as in steps 3 and 4.
12. BC = BJ and BD = AB, by construction as in step 5.
13. ΔCBD-ΔJBA, as in step 6 and highlighted in part (b) of the figure.
14. Rectangle BDFG = 2ΔCBD, as in step 8.
15. Square CKJB = 2ΔJBA, as in step 9.
16. Therefore, rectangle BDFG = square CKJB, as in step 10.
17. Square ABDE = rectangle AGFE + rectangle BDFG, by construction.
18. Therefore, square ABDE = square AIHC + square CKJB, by steps 10 and 16.

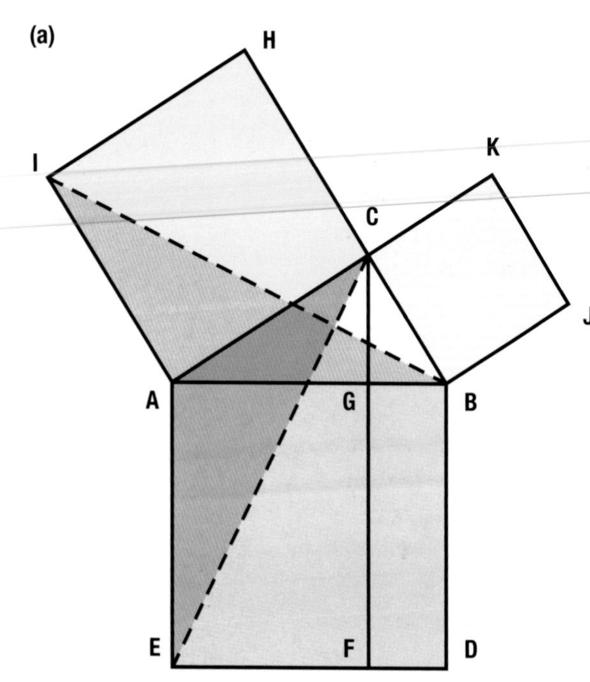

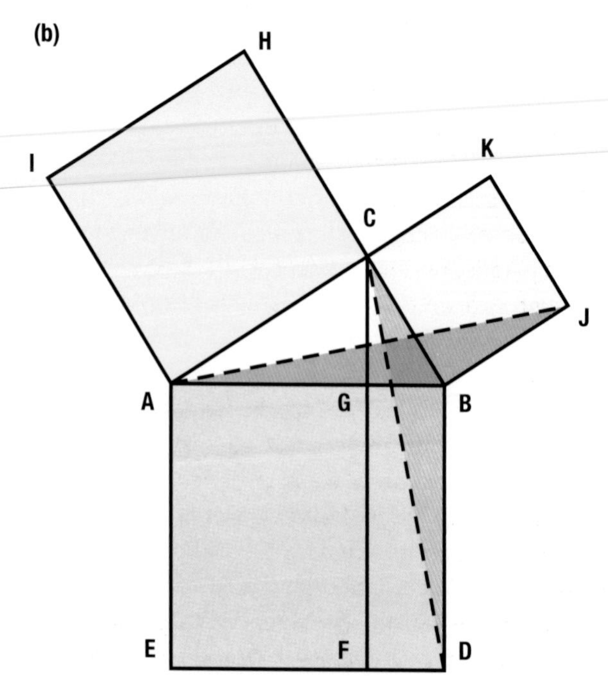

Postulates, theorems and proofs

Euclid and the Greeks gave mathematics extraordinary power by turning it into a logical system. They introduced the idea of proofs, and the idea that rules could be worked out logically from certain assumptions, or postulates, such as 'A straight line is the shortest distance between two points.' Assumptions are then combined to make a basic idea for a rule, called a theorem, which is then proved or disproved.

At the heart of Euclid's the *Elements* are five key postulates or axioms. In modern terms they are:

1. Part of a line can be drawn between two given points.
2. Such a part line can be extended indefinitely in either direction.
3. A circle can be drawn with any radius with any given point at its centre.
4. All right angles are equal.
5. If part of a line crosses two other lines so that the inner angles on the same side add up to less than two right angles, then the two lines it crosses must eventually meet.

The first four sound self-evident to us today, yet they were by no means self-evident to people of the time, and it was Euclid's efforts to define the most basic concepts that made his work so profoundly influential. Only with completely watertight definitions of the basic concepts is it possible to establish firm proofs of otherwise vague hunches. And only with completely watertight definitions can we move confidently and logically forward through each step – any looseness in the definitions immediately invalidates the chain of logic.

Parallel lines and Euclid's limitations

The fifth of Euclid's postulates is less self-evident, and is about parallel lines. If part of a line crosses two other lines so that the inner angles on the same side add up to exactly two right angles, then the two lines it crosses must be parallel. This fifth postulate is therefore called the parallel postulate. This postulate was held as a basic central truth, and it is at the heart of all basic geometric constructions and has countless practical applications: train lines, for instance.

However, Euclid was not entirely happy with his parallel postulate, and it turns out that he was right to have doubts. Euclid's geometry works perfectly for flat and two- or three-dimensional surfaces and most everyday situations. But just as the Earth's surface is not actually flat, however much it appears to be, so space is actually curved and has many more than three dimensions, including that of time. Euclid's parallel postulate means that only one line can be drawn parallel to another through a given point, but if space is curved and multidimensional, many other parallel lines can be drawn. Similarly, according to Euclid's geometry, the internal angles of a triangle always add up to 180 degrees – yet those of a triangle drawn on a ball add up to more than 180.

Such mathematicians as Carl Gauss began to realize the limitations of Euclidean geometry in the nineteenth century and to develop a new geometry for curved and multidimensional space. However, Euclid's work has been the bedrock of geometry for 2,200 years, and remains central to all everyday geometry today. Moreover, Euclid's method of establishing basic truths by watertight reasoning – that is, by logic, deductive reasoning, evidence and proof – is as powerful now as it ever was – so powerful that we take it as basic common sense.

Archimedes

Although Archimedes was amongst the world's most prolific inventors, he preferred to be remembered for his theories. His tomb was inscribed with a sphere and a cylinder, the discovery of the ratio of which was one of his proudest moments.

c. 287–212BC

'Give me a place to stand and I will move the Earth', Archimedes is said to have declared to King Heiron II of Syracuse in Sicily some time around 260BC. To the astonishment of those present, so the story goes, Archimedes had just singlehandedly launched the *Syracusia*, at 4,064 tons one of the biggest and most luxurious ships built in ancient times. Launching this beached monster had defeated all the efforts of huge teams of men pulling on ropes. Yet Archimedes, with an ingenious arrangement of levers and pulleys, performed the task by himself with ease.

No wonder then that he was a legend in his own lifetime and that tales of his genius spread far and wide. He was, without doubt, the greatest inventor of ancient times. Not only did he invent pulleys and levers to launch giant ships, but he built the first water pump, which is called an Archimedes screw and is still used in many places today. He created a wonderful planetarium to show the motions of all the planets and invented a machine to fire burning pitch at enemy ships. And when his home city of Syracuse was besieged by a Roman fleet, he constructed catapults to bombard the ships with boulders, a mirror to focus sunlight and set them on fire, grappling devices to throw down scaling ladders, and even a hook and crane to lift the huge enemy boats right up out of the water and tip them over.

Yet in some ways Archimedes's inventions are the least of his achievements. Indeed, he himself did not hold them in much esteem. Like most Greek thinkers, he placed a higher value on abstract scientific and mathematical ideas than on their practical applications. The Roman writer Plutarch insisted that Archimedes:

> *did not deign to leave behind any written work [on practical inventions]; he regarded as sordid and ignoble the construction of instruments, and in general every art directed to use and profit, and he strove after those things which, in their beauty and excellence, remain beyond all contact with the common needs of life.*

All we know of Archimedes suggests that Plutarch was seriously overstating the case, because Archimedes, more than any other thinker of his time, did not hesitate to build machines to try out his ideas, and to construct practical experiments. And he was genuinely excited by his own inventiveness. All the same, it is his purely intellectual achievements that were his lasting legacy and made him the greatest scientist in history until the time of Isaac Newton, who was in awe of him.

In fact, Archimedes was the world's first great scientist. Other great minds before him had studied scientific subjects, and there are many less famous Greek thinkers whose achievements in science deserve to be remembered, but he was the first person to think about problems in the scientific way we now take for granted. His abstract theories could all be proved or disproved by practical experiments and mathematical calculations, which is the method that has led to nearly every achievement in science to this day.

Archimedes's life

Archimedes was born in 287BC in Syracuse in Sicily, which was then a Greek colony. He was Greek, not Sicilian. The city was a frontier town, sandwiched between the warring powers of Rome and Carthage, and yet it was by no means an intellectual backwater. King Heiron II and his son King Gelon were enlightened, intellectually inclined rulers. Indeed, Archimedes may have been Gelon's tutor.

All the same, if anyone wanted a proper education, Alexandria in Egypt was the place to go, and here Archimedes went as a young man. At the time he was there, the city was the greatest centre of learning in the ancient world. Although the museum or university there was barely 20 years old – the city itself had been founded by Alexander the Great just half a century earlier – it already held an unrivalled library, containing at least 100,000 scrolls, including all of Aristotle's priceless personal collection. It was here that the great Euclid taught geometry, that Aristarchus showed that the Earth revolved around the Sun, and that Hipparchus made the first great catalogue of constellations, categorizing stars in terms of their brightness. And it was here that, much later, Ptolemy wrote the *Almagest*, the most influential book about the nature of the universe for 1,500 years. Euclid was probably dead by the time Archimedes was there, but Archimedes undoubtedly met Eratosthenes, the brilliant thinker who measured the circumference of the world to within 4 per cent of modern figures, and made a measurement of the year's length as precise as any until barely half a century ago.

On noticing that the water level in the bath rose proportionately to the size of the body immersed in it, Archimedes leapt naked from his bath and ran down the street shouting 'Eureka!'

Although Archimedes got his grounding in science and mathematics in Alexandria, his activity there was by no means purely intellectual. Some reports say he was employed for a time as an engineer on large-scale irrigation works in the Nile Delta, and it was probably while he was in Egypt that he invented his famous screw for pumping water.

But once he had returned to Syracuse, he stayed there all his long life – inventing, studying, thinking. Accounts of his life in Syracuse paint him as the archetypal absent-minded scientist, so absorbed in high thoughts that he neglected everyday needs.

The most famous story about him concerns a discovery he made whilst in the bath. King Heiron had given a goldsmith some gold and asked him to make a wreath from it. When the goldsmith finished the wreath, Heiron suspected the crafty goldsmith had pocketed some of the gold and replaced it with some cheaper metal. Yet the wreath weighed exactly the same as the original gold. How could the fraud be proved. Heiron asked Archimedes, and even he found it a tricky problem. Then one day, while mulling over the problem in the bath, he suddenly noticed how the water level rose as he sunk deeper into the bath.

The story goes that Archimedes leaped straight out of his bath and ran naked through the streets to the king, shouting at the top of his voice, *'Eureka! Eureka!'* (I've got it! I've got it!). Later he showed the king his idea. First he immersed in water a piece of gold that weighed the same as the wreath and pointed out the subsequent rise in the water level. He then immersed the wreath itself and showed that the water level was higher than before. Archimedes said that this meant that the wreath must be a greater volume than the gold, even if it was the same weight. Therefore it could not be pure gold. The fraudulent goldsmith was executed.

Whether this story is true or not, it is typical of Archimedes's amazingly neat and elegant scientific solutions to awkward questions – and of how a small practical problem led him on to crucial theoretical insights. It may be that this was the starting point for his groundbreaking work in hydrostatics – how things float (see box on p16).

Mathematical insights

Archimedes also tried to approach problems mathematically. He may not have been the first to realize that if a weight is put on each end of a seesaw, the lighter weight must be further away than the heavier if the two weights are to balance. But Archimedes went further and showed that the ratio of weights to one another goes down in exact mathematical proportion to the distance from the pivot of the seesaw – and proved it mathematically. In the same way, he had the brilliant insight that every object has a centre of gravity – a single point of balance from which all its weight seems to hang – and again proved it mathematically.

Interestingly, besides looking at practical problems in a mathematical way, he also looked at mathematical problems in a practical way, and this, if anything, was even more revolutionary – though it took over 2,000 years for others to realize it. The work that Archimedes was proudest of was his solutions to geometric problems – especially the problems of working out the volumes and areas of regular shapes, such as spheres and cones.

Some of his mathematical achievements were purely abstract in the Greek tradition. He showed, for instance, that the surface area of a sphere is four times the area of its 'greatest circle' – in other words four times the area of a circle with the same radius. He also showed that the volume of a sphere is two-thirds of the cylinder into which it fits perfectly. Indeed, he was so proud of this discovery that he asked for a diagram of a sphere inside a cylinder to be inscribed on his tomb, and it was.

But it was when he introduced practical ways of working that he achieved the greatest

The Sand Reckoner

In a famous letter, now called *The Sand Reckoner*, written to his protégé King Gelon, Archimedes showed that mathematics is capable of dealing with unimaginably large numbers. He wrote: 'There are some, King Gelon, who think that the number of grains of sand is infinite in multitude.... But I will show you by mathematical proofs that ... some numbers exceed [the number of grains in an entire universe filled with sand].' Archimedes showed that by building up numbers in levels, now called powers, it was possible to create gigantic numbers. 2 times 2 is 2 to the power of 2, that is 4. 2 times 2 times 2 is 2 to the power of 3, that is 8. 2 times 2 times 2 times 2 is 2 to the power of 4, that is 16. Archimedes quoted the number P to the power of 100 million, which is a pretty huge number, especially if P itself is large, but implied that one could go on even further.

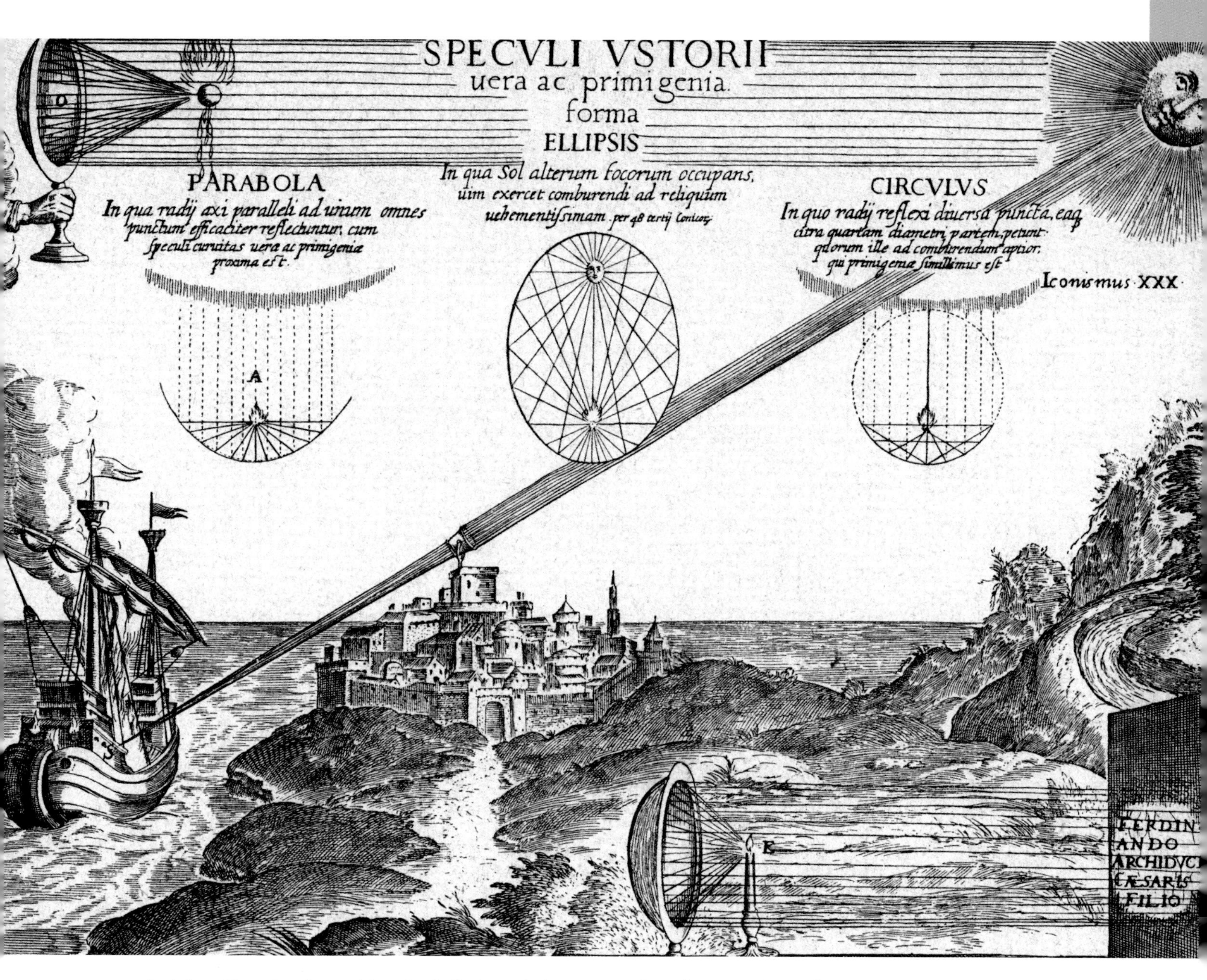

An illustration showing sun rays and a mirror, purportedly developed by Archimedes in 214BC, and used to set fire to Roman ships at Syracuse during the Second Punic War between Rome and Carthage.

insights. The Greek tradition had disdained the practical. Following Plato, the Greeks believed pure mathematics was the key to the perfect truth that lay behind the imperfect real world, so that anything that could not be completely worked out with a ruler and compass and elegant calculations was not true. Archimedes's genius was to see through the limitations of this and realize just how much could be achieved through practical approximations, or, as the Greeks called them, mechanics. It is clear he knew how much he was working against the Greek tradition when he wrote to a colleague in Alexandria about his solution to a problem: 'This concerns a geometrical theorem which has not been investigated before but has now been investigated by me. I first discovered this theorem by means of mechanics and then demonstrated it by means of geometry.'

This approach certainly bore fruits for Archimedes. For instance, he was able to work out the approximate area of a circle by first working out the area of the biggest hexagon that would fit inside it and then the area of the smallest that would fit around it with the idea in mind that the area of the circle must be pretty much halfway in between. By going from a hexagon to polygons with ninety-six sides, he

Floating and sinking

Archimedes single-handedly launches the ship Syracusia, *using only an arrangement of pulleys and levers.*

One of Archimedes's greatest discoveries was that of buoyancy and just why things float. He realized that an object weighs less in water than in air. A quite heavy person can float in a swimming pool because of his buoyancy – the natural upward push or upthrust of the water on him. However, when an object is immersed in water, its weight pulls it down. But the water, as Archimedes realized, pushes back up with a force that is equal to the weight of water the object pushes out of the way. So the object sinks until its weight is exactly equal to the upthrust of the water, at which point it floats. Objects that weigh less than the water displaced will float and those that weigh more will sink. Archimedes showed this to be a precise and easily calculated mathematical relationship.

This was a very, very important breakthrough, because it allowed shipbuilders to predict whether their ships would float, rather than just proceeding by trial and, all too often, catastrophic error. But Archimedes went further, making calculations about how all kinds of different shapes would float, and the angle they would float at it. Even though much of his work was purely theoretical, it had crucial practical implications – for it enabled shipbuilders to work out whether or not ships would tip over.

could narrow the margin for error considerably – and so found a figure for pi, 22/7, that is still good enough for most mathematical calculations today.

In the same way, he worked out the approximate area contained by all kinds of different curves from the area of rectangles fitted into the curve. The smaller and more numerous the rectangles, the closer to the right figure he got. This is the basis of what thousands of years later came to be called integral calculus, which became perhaps the most powerful of all mathematical tools for science in the hands of scientists like Newton and his successors.

Archimedes meets his death in Syracuse at the hands of a non-mathematically inclined Roman soldier.

Death and legacy

By the time the Roman fleets besieged Syracuse in 212BC, Archimedes was an old man – perhaps nearly 80 – but he was still actively inventive and busy at work on theoretical problems. As the Roman ships closed in on the city, the ageing Archimedes was in the thick of things, coming up with all kinds of ingenious contraptions to keep the enemy at bay. But even the genius of Archimedes could not keep the Romans at bay forever.

The Roman commander Marcellus had been so impressed by Archimedes's contraptions that he insisted the scientist should be treated well when his troops finally entered the city. Sadly, the Roman officer who came upon Archimedes had not got the message. According to one story, the soldier burst in through his door and found him at work drawing circles and making calculations in his sand tray. 'Please do disturb my calculations,' he barked. The battle-weary soldier was in no mood for argument and ordered Archimedes to come with him. Archimedes insisted on finishing his calculations first – and so the soldier drew his sword and killed him.

Sadly, despite his fame, much of Archimedes's work was lost and was only gradually recovered. Some important works have never been found. Remarkably, one major work was only rediscovered in 1906, when the Danish philologist J. L. Heiberg found that a medieval parchment discovered in a Jerusalem monastery was a palimpsest – that is, a scroll in which the original writing has been partially erased so that it can be used for a new text. Here, beneath Greek orthodox scriptures, were hidden copies of various key works by Archimedes.

Enough of Archimedes's work was kept alive by Arab mathematicians through the Dark Ages for it finally to be taken of advantge of when the scientific revolution began in Europe in the seventeenth century. Galileo acknowledged his debt to the remarkable Greek scientist by saying, 'Without Archimedes, I could have achieved nothing', while Newton said, 'If I have seen further, it is only by standing on the shoulders of giants' – and for him the greatest of all giants was Archimedes.

Hipparchus & Claudius Ptolemy

Hipparchus (c. 200–126BC) and Ptolemy (c. 90–168AD) were the greatest astronomers of the ancient world. Together their work formed the basis of astronomy for the next 1500 years.

c. 200–126BC
c. 90–168AD

Hipparchus and Ptolemy compiled the first comprehensive catalogues of the stars in the sky, constructed systems for working out the motion of the Sun and Moon, and much more besides. Astronomy dates back to the very earliest days of humankind, when prehistoric hunters gazed up the sky to work out which night might give them the best moon for hunting. When people began to farm about 10,000 years ago, astronomy helped pinpoint the best time to plant crops. It became so important that astronomers often held the status of high priests, and many important ancient monuments, such as the pyramids and Stonehenge, have strong links with astronomy.

So by the time Hipparchus was born in about 200BC, astronomy was already an ancient art. Very little is known about Hipparchus, even though he was famous enough to be depicted on Roman coins after his death. He was probably born in Nicaea in Bithynia, now northwestern Turkey, by Lake Iznik. It is likely that as a young man he compiled records of local weather, to try and link the timings of seasonal weather patterns with the rising and setting of particular stars. But most of his life was probably spent studying astronomy on the island of Rhodes, after a short spell in Alexandria in Egypt. Ptolemy claimed that Hipparchus made many observations of the stars from Rhodes. He died in Rhodes, perhaps around 126BC. Our knowledge of his work is inferred rather than certain.

Hipparchus's astronomical writings were so extensive that he published an annotated list of them, yet only one work of his has survived, a short commentary on a popular poem called *Phenomena*, by Aratus and Eudoxus, which describes the constellations. Although it tells us nothing about his astronomy, it shows something of his rigorous, critical attitude to loose thinking as he ruthlessly points out the errors in the poem's description of the stars. Ptolemy described him as a 'lover of truth', and if he could be critical and judgemental, he was harshest with his own work, always ready to revise his ideas if new evidence came along.

Hipparchus at work

Hipparchus was undoubtedly a skilled observer and saw many things anew for himself, but he also drew on the long history of astronomy in the Middle East, and in particular ancient Babylonians records rescued from the ruins of the Persian Empire after Alexander's conquest.

In 134BC, he spotted that rare phenomena, a new star, or nova, in the night sky – there was not to be another such sighting until Tycho Brahe spotted one in 1572. It is said that this extraordinary event was what inspired Hipparchus to compile a catalogue of all the 850 or so stars whose positions were then known. This catalogue, adapted by Ptolemy, was still in use in the sixteenth century. Indeed, it was so accurate that 1,800 years on, Edmund Halley was able to compare his own map with Hipparchus's catalogue and see that stars change their positions slightly over the centuries.

Hipparchus also compared stars by giving each one a magnitude number from one to six, depending on how bright it looked. The brightest star is Sirius (the Dog Star), which he called a First Magnitude star. The faintest stars he called Sixth Magnitude. Although the scale has been adapted and extended, astronomers still use it today.

What made Hipparchus such a great astronomer was his precision. Considering he had only his own naked eyes and vague historical records to guide him, he made some astonishingly precise calculations of the movements of the heavens. We sometimes mistakenly think that the ancients had no real knowledge of where the Earth sits in the solar system – or even that it was round and not flat – but Hipparchus (and other Greek astronomers) did have some idea, although it was not accurate.

Hipparchus's only large error was to assume, like all those of his time except for Aristarchus of Samos, that the Earth is stationary and that the Sun, Moon, planets and stars revolve around it. But the fact that the stars are fixed and the Earth is moving makes such a tiny difference to the way the Sun, and Moon and stars seem to move that Hipparchus was still able to make highly accurate calculations of their movements.

Hipparchus's calculations

At the heart of Hipparchus's astronomical precision was his mathematical skill and rigour. It is said that he invented the branch of mathematics called trigonometry – the mathematics of triangles – and developed chords, the first trigonometrical tables, which helped him calculate the precise position of a star in the sky relative to the Earth and other stars.

Some of Hipparchus's most important astronomical calculations came from plotting the the ecliptic, which is the circular path of the Sun through the sky. The ecliptic is at an angle to the Earth's equator, and crosses it at two points, the equinoxes, and takes the Sun furthest away from it at the solstices. What intrigued Hipparchus was that even though the Sun apparently travelled a circular path, the seasons – the time between the solstices and equinoxes – were not of equal length. To pin this down, he worked out a method of calculating the Sun's path which would show its exact location on any date.

He went on to measure as precisely as possible the length of a year. There are several ways of measuring a year astronomically. Hipparchus measured the 'tropical year', the time between equinoxes. The results from his own observations were inconclusive, but he could correct any of his own errors by comparing them with ancient records, and thus he arrived at a figure for the year length that was just 6 minutes too long.

The wobbling Earth

It was from this amazingly accurate observation that came what is perhaps Hipparchus's most famous discovery – the idea of the 'precession of the equinoxes'. When he calculated the exact timing and position of the stars at the equinox and compared it with observations taken 150 years earlier, he discovered that stars near the ecliptic had moved their positions slightly. After considering various explanations, he concluded that the whole star pattern was moving slowly eastwards – and that it would go round and return to the same position every 26,000 years. We know now that this movement is caused by a slow change in the direction of the Earth's tilt, called precession, rather than a shift in the stars, but Hipparchus was basically right, and it was a remarkable observation.

From this, Hipparchus went on to study the motion of the Moon in order to work out when and why eclipses occurred. Apparently he only made a little progress with this problem because he would not accept a theory that could not be absolutely verified by his observations. What he did do, however, was measure just how far away the Moon is compared to the size of the Earth. To do this he relied on the phenomenon of parallax. As the Earth moves – or, to Hipparchus, as the sky moves – objects in the sky nearby appear to move slightly sideways compared to those further away. The more an object moves, the nearer it must be. Using a complex but ingenious series of assumptions, he compared the size of the Moon during solar eclipses (the same apparently as the Sun) and the Earth's shadow on the Moon

The Harmonia Macrocosmica: a print by Andea Cellario showing the Ptolemaic system. This system postulated a geocentric solar system, in which the Earth was at the centre of the universe, a system that was accepted for over 1000 years.

Ptolemy

We know of Hipparchus's work partly because it was developed by the astronomer Claudius Ptolemy (AD90–168), who wrote four books summarizing Greek astronomical ideas in the second century AD, including the famous *Almagest* (Arabic for 'The Greatest'). These books, and in particular the *Almagest*, became the cornerstone of Western and Arab astronomy until the sixteenth century.

Ptolemy as a person is even more obscure than Hipparchus. We know he was Greek and lived in Alexandria, but that is about all. However, his major works survived and proved to be hugely influential.

The *Almagest* provided a complete system for the movements of the heavens that came to be called the Ptolemaic system. It provided the basis for all astronomy until it was finally overturned by the Copernicans in the sixteenth century (see p. 33). At the centre of the system is the fixed Earth. Around it rotates a vast sphere carrying with it in a series of layers, the stars, planets, Sun and Moon – and this explains their movement through the sky and their rising and setting.

As we now know, the planets do not appear to follow a perfectly circular route through the sky because of the movement of the Earth. Instead, they appear to loop back on themselves every now and then, earning them their name, which is Greek for 'wanderers'. Ptolemy's system explained this ingeniously with a system of circles within circles called epicycles, rather like some vast celestial clockwork mechanism that continually whirrs round. What was remarkable was how accurately it worked, allowing astronomers to make very precise predictions about where stars and planets would be. No wonder, then, even when Copernicus pointed out the fatal flaw – that the Earth moves – astronomers were reluctant to abandon it.

Ptolemy's book *Geography* was, if anything, more influential than his *Almagest*. It was a collection of maps of the whole world as his contemporaries knew it. His great innovation was to record longitudes and latitudes in degrees for 8,000 locations on his world map. He also devised two ways of drawing grid lines on flat maps to represent the lines of longitude and latitude on the curved surface of the globe. Of course, Ptolemy's knowledge of the world was no greater than that of any of his contemporaries, and despite all the precision he could muster, the maps were as wildly inaccurate as the information he was working with. Nevertheless, they were as good as anyone could make at the time, and *Geography* became the standard atlas for another 1,300 years. Indeed, Christopher Columbus is said to have believed he could reach Asia easily by sailing west across the Atlantic simply because Ptolemy hugely underestimated the size of the world – and so Columbus accidentally discovered the Americas while looking for Asia in the west.

during lunar eclipses. In this way, he worked out that the Moon is about sixty-three times the Earth's radius away. It is in fact a little over sixty.

We know about most of Hipparchus's work only second-hand, as it appears in Ptolemy, for instance. But in January 2005, a real piece of hard evidence was found to show that Hipparchus's reputation is not unfounded. Previously, the evidence that Hipparchus had really compiled an accurate star catalogue was vague, but then American astronomical historian, Bradley Schaefer, began to examine a 2.13 m tall Roman marble statue of the mythical titan Atlas in the Farnese Collection in Naples, Italy.

The statue carries a globe which shows all the constellations in exactly the right locations, as if they were taken from a star catalogue. By analyzing the positions of the stars on the globe, Schaefer calculated that the observations which dictated where the stars would appear on the globe must have been made around 125BC, give or take 55 years. This is exactly the time Hipparchus was working, and so provides strong evidence that he did produce a star catalogue. In the future, experts may compare the stars in the Farnese Atlas with those in Ptolemy's *Almagest* and see just how different they are. In the meantime, it indicates that Hipparchus's reputation as the greatest astronomer of the ancient world is not misplaced.

The Medieval Arab Scientists

During the period known in Europe as the Dark Ages, the Middle East became the intellectual hub of the world, and a string of brilliant Arab scientists sowed the seeds that would later be reaped in the scientific revolution of the seventeenth century.

fl. 8th century AD

AL-KHWARIZMI GAVE US OUR NUMBER SYSTEM AND ALGEBRA, Ibn Sina (Avicenna) wrote the greatest book on medicine for a thousand years, and al-Biruni's pioneering work included comparisons of the speed of light and sound. This great flowering of Arab science began in what is sometimes called the Golden Age of Islam, the time when the consolidating influence of Islamic religion saw Arab Muslims begin to build an empire that was to stretch across the Middle East and across North Africa into Spain. At the heart of this Islamic world was the city of Baghdad, where the Caliphs ruled.

The rule of the Caliphs reached its zenith under the Abbasid dynasty, and in particular in the reign of the Caliph Harun al-Rashid and his son al-Mamun. Harun al-Rashid, who became caliph in AD786, has become famous for the fictional role given him in *The Thousand and One Nights*, which may have been partly written in his reign. But his reign marked the beginning of an astonishing flowering of culture in Baghdad. The nature of Islam meant that scholarship was treasured, and Baghdad became a centre of unrivalled intellectual activity in all fields, including science, technology, poetry and philosophy. The Arab scholars saw no distinction between these branches of thought, and many would study mathematics or write poetry with equal zeal.

Under the caliphate of Harun's son al-Mamun from 813 on, Baghdad drew scholars from far and wide to work at the House of Wisdom, created by the caliph about 820. The House of Wisdom was a mixture of library, research institute and university. Indeed it was the first great library since the library at Alexandria had been destroyed, possibly in the first century BC. One of the House of Wisdom scholars' tasks was to translate the great works of the Greek thinkers, and it is largely through their efforts that Greek ideas were preserved through the Dark Ages.

But the scholars also did practical research, establishing the world's first proper astronomical observatory, for instance. They also developed the astrolabe, one of the most influential scientific instruments of all time, allowing astronomers to measure the position of stars with unparalleled accuracy. In medicine, they improved diets, made the first serious studies of drugs, and advanced surgery hugely. Indeed, Muslim scholars of the Golden Age made important and original contributions to mathematics, astronomy, medicine and chemistry.

When al-Mamun died in 833, the central role and influence of Baghdad began to wane. But in its place, energetic pockets of scholarship began to spring up throughout the Islamic world. The great al-Biruni lived under the patronage of the Ghaznavid caliphs in the east while Avicenna lived under the Sammarid caliphs in Bukhara.

Avicenna's treatment for a fracture-dislocation to the spinal column. TOP: pressure is applied using leverage on a board; MIDDLE: pressure is applied by pounding with a heavy instrument; BOTTOM: the patient is in traction and experiencing local manipulation.

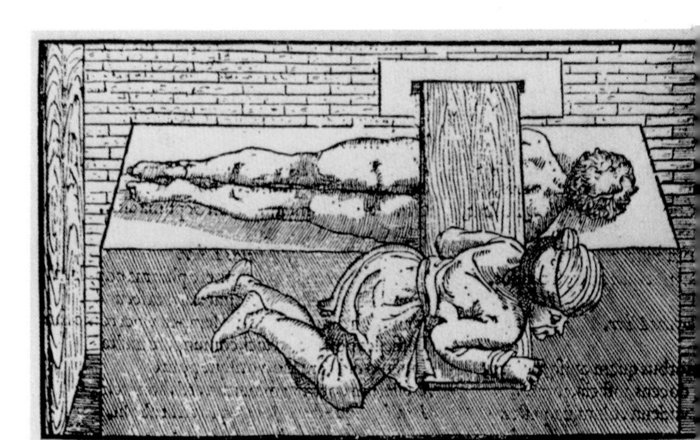

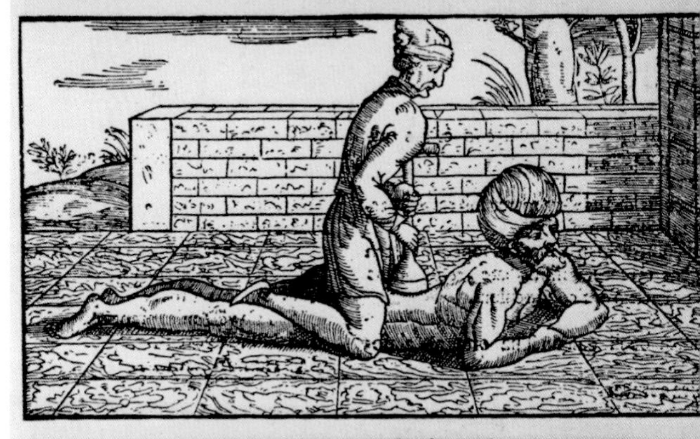

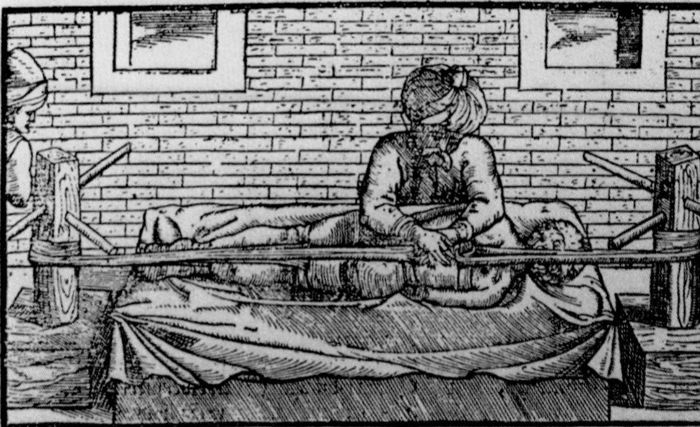

Al-Khwarizmi

Very little is known about the life of al-Kharwizmi, and much of his life story seems to be pure conjecture. It seems likely that he was born in what is now Uzbekhistan, south of the Aral Sea in central Asia. Some scholars say his father was a Zoroastrian and that he was brought up in this ancient faith, which dates back to the time of ancient Sumeria. Others say this is to completely misinterpret the records. All we do know is that al-Khwarizmi was born about 786, the year Harun al-Rashid came to power, and that when Harun's son al-Mamun set up the House of Wisdom, al-Khwarizmi was there studying.

One of his greatest contributions was to provide a comprehensive guide to the Hindu numbering system which originated in India in about 500AD. It is this system, later called the Arab system because it came to Europe from al-Khwarizmi, that became the basis for our modern numbers. The Arab system, clearly explained by al-Khwarizmi in his book *On Hindu Numerals*, uses only ten digits from 0 to 9 to give every single number from zero up to the biggest number imaginable. The value given to each digit varies simply according to its position. So the 1 in 100 is 10 times the 1 in 10 and 100 times the 1 in 1.

It was a huge advance on every previous numbering system, which became incredibly cumbersome with any large numbers. The Roman numbering system, for instance, needs seven digits to give a number as small as, for example, 38 – XXXVIII. Arabic numbering can give even very large numbers quite compactly. Seven digits in Arabic numerals can, of course, be anything up to ten million. What's more, by standardizing units, Arabic numerals made multiplication, division and every other form of mathematical calculation much, much simpler. Not surprisingly, when it reached Europe, it quickly caught on, and has since spread around the world to become the one truly global 'language'. With the numbers, Europe gained another word, too – "algorithm" for a logical step-by-step mathematical process, based on the spelling of al-Kharizmi's name in the Latin title of the book, *Algoritmi de numero Indorum*.

Inventing algebra

If the introduction of Arabic numerals owed something to Hindu mathematicians, al-Khwarizmi's other major contribution was much more his own work. This, too, introduced a word to our language, 'algebra', and a whole new branch of mathematics. What is interesting is that in developing algebra, al-Khwarizmi had

Ibn Sina (Avicenna)

Born in Bukhara around 980, Ibn Sina – sometimes known by his Latin name Avicenna – was something of a prodigy. It is said that by the age of 10 he knew not just the Koran but most Arabic poetry by heart, and by the age of 16 had become a qualified physician. His skill as a physician became almost legendary, and although the turbulent politics of the time kept him permanently unsettled, he was physician to a succession of princes and caliphs. But he also became the most famous philosopher, scholar, mathematician and astronomer of the age, and wrote books on a huge range of scientific topics, a vast encyclopedia – the first ever written – and even wrote short poems, one of which goes:

"Up from Earth's Centre through the Seventh gate I rose,
And on the Throne of Saturn sate
And many a knot unravelled by the Road
But not the master-knot of Human Fate."

Ib Sina made a number of key astronomical observations, devised a vernier scale to help make readings more precise, and made a string of key contributions to physics such as identifying the different forms of energy – heat, light and mechanical – and the idea of a force. He noted that if light consists of a stream of particles, then its speed must be finite. The mathematical technique of 'casting out of nines', used to verify squares and cubes, is also attributed to Ibn Sina.

Yet his fame is above all based on his book *al-Qann fi al-Tibb* (*The Canon of Medicine*). This vast tome, consisting of over a million words, surveyed the entire field of medical knowledge from ancient times to the most up-to-date Muslim techniques. Its comprehensive, systematic approach meant that once it was translated into Latin in the twelfth century it became the standard medical textbook in Europe for the next 600 years.

Besides bringing together existing knowledge, the *Canon* contained some of Ibn Sina's own insights. He was the first to recognize, for instance, that tuberculosis and phthisis are contragious; that diseases can spread through soil and water; and that a person's emotions can affect their state of physical health. He was also the first to describe meningitis and realize that nerves transmit pain. The book also contained a description of 760 drugs, and so became the doctor's bible for centuries to come.

something eminently practical in mind, not just abstract theory. In his introduction to the book in which he describes algebra, he says the aim is to work with 'what is easiest and most useful in mathematics, such as men constantly require in cases of inheritance, legacies, partition, lawsuits, and trade, and in all their dealings with one another, or when measuring lands, digging canals and making geometrical calculations.'

Although we now associate algebra entirely with the idea of symbols replacing unknown numbers in calculations, al-Khwarizmi did not actually use symbols, for he wrote everything out fully in words. It was in his way of handling equations that he created algebra.

Completing and balancing

In his work on alegbra, Al-Khwarizmi worked with both what we now call linear equations – that is, equations that involve only units without any squared figures – and quadratic equations, which involve squares and square roots. His great breakthrough was to reduce every equation to its simplest possible form by a combination of two processes: *al-jabr* and *al-muqabala*.

Al-jabr means 'completion' and involves simply taking away all negative terms. Using modern symbols, *al-jabr* means simplifying, for instance, $x^2 = 40x - 4x^2$ to just $5x^2 = 40x$. *Al-muqabala* means 'balancing', and involves reducing all the positive terms to their simplest form. *Al-muqabala* reduces, for instance, $50 + 3x + x^2 = 29 + 10x$ to just $21 + x^2 = 7x$.

In this way, he could reduce every equation to six simple, standard forms, and then showed a method of solving each. He went on to provide geometrical proofs for each of his methods. Some scholars say that these prove he must have read Euclid's geometry, especially as some of his colleagues in the House of Wisdom had translated Euclid's the *Elements*. Others say they are so markedly different from Euclid that his geometry came from elsewhere.

In developing algebra, al-Khwarizmi built on the work of Hindu mathematicians such as Brahmagupta, but it was al-Khwarizmi who turned it into a simple, all-embracing system, and he is rightly dubbed the 'father of algebra'. The very word algebra comes from the title of his book *al-Kitab al-mukhtasar fi hisab al-jabr-wa'l muqabala* or *The Compendious Book on Calculating by Completion and Balancing*.

The world and the stars

Like all the Arab scholars, al-Khwarizmi had interests beyond mathematics. Apart from his books on numbers and algebra, he wrote a third major book, on geography. Called *Kitab surat al-ard* (*The Image of the Earth*) this reintroduced some of Ptolemy's ideas on describing positions through longitude and latitude, but improved on their accuracy – particularly for the length of the Mediterranean Sea and the location of cities in Asia and Africa. Al-Khwarizmi also helped to create a map of the world for the Caliph al-Mamun, and got involved in a project to measure the circumference of the Earth by measuring the length of a degree of longitude through the plain of Sinjar in Iraq. Scholars in this part of the world had long moved beyond the idea that the world is flat.

At some time in his life, al-Khwarizmi worked on astronomical problems too. He compiled astronomical tables, and developed a table of sines for helping calculate the position of stars.

Al-Khwarizmi's work took perhaps three centuries to reach western Europe, and probably came via Moorish Spain. But when it did arrive, its impact was lasting. Arabic numbers and algebra are central to our lives today.

An Arabic astrolabe from about the fourteenth century. An astrolabe is an instrument used to reckon how the sky appears at a given time and place.

al-Biruni

The Persian scholar al-Biruni lived around the same time as Ibn Sina and the two are known to have corresponded. He had a special gift for languages and is said to have spoken Turkish, Persian, Sanskrit, Hebrew and Syriac besides his native Arabic. He was patronized by the Ghaznavid caliph Mahmud, who took al-Biruni with him when he went on his campaigns to India. This is probably how al-Biruni learned the languages. Certainly his most famous book, *Kitab al-Hind* (*Book of India*) came out of these travels. Although he is not known for any particular scientific advance, his contributions ranged across a wide spectrum. He was the first to firmly establish trigonometry as a branch of mathematics. He wrote treatises on the effect of drugs in medicine. He pioneered the idea that light travels faster than sound. He promoted the idea that the Earth rotates on its axis, made many accurate calculations of latitude and longitude, and suggested, contrary to received opinion at the time, that Africa might not stretch on south forever. He also accurately measured the density of 18 precious stones and metals, and noted that flowers have 3, 4, 5, 6 or 8 petals, but never 7 or 9.

Leonardo da Vinci

Although he is best known for his masterful paintings, such as the Mona Lisa, Leonardo da Vinci (1452–1519) is revealed as a remarkable scientist by his notebooks, perhaps the first great scientist of the modern age.

1452–1519

We know that Leonardo was a remarkable scientist only from the surviving pages of his notebooks. There are thousands of these pages, each packed with Leonardo's extraordinary drawings and writings on a huge range of subjects – geology, anatomy, astronomy, gravity, flight, optics and much, much more. Often Leonardo flits from one topic to another on a single page.

Most famous of all are Leonardo's inventions. Within the pages are sketches and notes for scores of machines and devices, some just tentative ideas, some fully worked out with detailed drawings. What is astonishing is not just the sheer range of problems that Leonardo put his mind to, from war machines to water supply, but just how many of his ideas are almost unnervingly ahead of their time. Helicopters, tanks, cars, aeroplanes, bicycles, parachutes – all appear in Leonardo's pages, 500 years before they became a reality. It seems unlikely, though, that Leonardo tried out many of these amazing ideas. It seems unlikely, too, that anyone else even knew about them except the few who acquired pages of the notebooks down the ages. The same goes for his scientific writings.

Leonardo's tiny writing – often written backwards in mirror form, probably to make it easier for him to write left-handed – is difficult to decipher. But as a result of scholars' studies of Leonardo's scientific writings, he has emerged as one of the finest scientific minds in history. His notes on subjects from anatomy to astronomy reveal him to have been almost as far ahead of his time in his science as in his inventions. In his geology in particular, Leonardo was discussing sediments, strata, fossils and the age of the Earth in a way that anticipated the great debates of the early nineteenth century, over 300 years later. Even more significantly, Leonardo's great emphasis on first-hand observation foreshadows the scientific approach that was to have such an impact centuries later. He wrote, 'Things of the mind left untested by the senses are useless'.

Yet, for whatever reason, Leonardo kept his thoughts to himself. No one knows exactly why he wrote the notebooks. Most people believe his plan was eventually to publish them as a book. But by keeping quiet about his ideas, he ensured that for all his extraordinary insights, he actually had little impact on the progress of science, and is best known today as an artist. One can only speculate how different things might have been if his ideas had become known.

Childhood in Vinci

Leonardo was born on 15 April 1452 in the little Tuscan town of Vinci. His mother was a 16-year-old servant girl called Caterina, while his father was a local notary (solicitor) called Ser Piero. After the birth, Ser Piero married a local heiress called Albiera, while Caterina was quickly married off to a local cowherd, leaving the baby in Ser Piero's care. Ser Piero and his new wife had little time for the infant, who was looked after mainly by his grandparents and his uncle Francesco.

Even as a child, Leonardo proved to be extraordinarily gifted. He learned to play the lyre and sing beautifully, was adept at horse riding, and showed a remarkable aptitude for mathematics. He would often go wandering through the countryside by himself around Vinci, always carrying a notebook to make his already striking sketches of plants and animals.

Leonardo in Florence

When his grandfather died in 1468, the family moved to Florence. Florence at the time was the most exciting, creative city in Europe. Dominating it all was Brunelleschi's spectacular new cathedral dome, but down below in the teeming streets were scores of workshops and studios turning out a stream of brilliant art, such as Ghiberti's Baptistry doors and Donatello's statue of David. Leonardo's father was by now well aware of his son's artistic talents and sent him to study at the studio of Andrea Verrochio, then the most famous sculptor, painter and goldsmith in Florence.

Leonardo was a fast learner, and soon surpassed his master in skill, causing Verrochio, legend has it, to give up painting in despair. At this time, he cut a dashing figure around town, wearing over-short, shocking pink breeches. People have speculated about his sexuality, and he seems never to have had any interest in women, writing in his notebooks, 'The act of procreation and anything that has any relation to it is so disgusting that human beings would soon die out if there were no pretty faces and sensuous dispositions.' Whatever the truth, his disinterest in personal relationships left him with plenty of time to work on his ideas and develop his art.

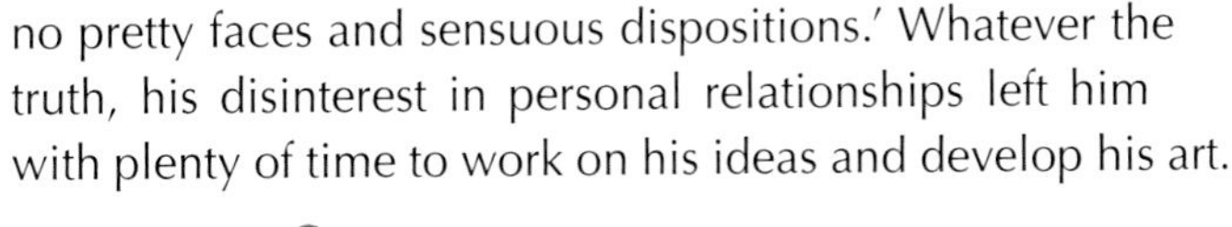

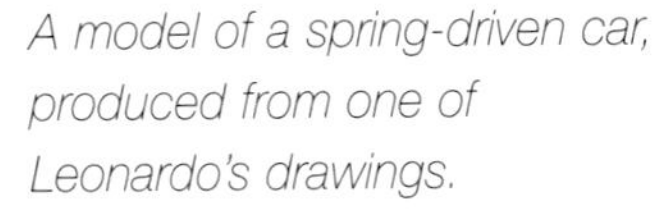

A model of a spring-driven car, produced from one of Leonardo's drawings.

Leonardo in Milan

In 1480 he received his first great art commission from Lorenzo the Magnifi-cent, the head of Florence's ruling family, the Medicis. Leonardo began work on this painting, *The Adoration of the Magi*, but before long abandoned it to write to Ludovico Sforza, the Duke of Milan, advertising his services, not primarily as a painter and sculptor, but as a military engineer. In his letter of introduction, he talked of his ability to make armoured cars and siege engines, portable assault bridges and catapults. Ludovico ignored this letter, but later summoned Leonardo to Milan, where he stayed 17 years until the French captured the city in 1499 and Ludovico fled.

Throughout his time in Milan, Leonardo was kept busy painting, staging court festivals, and advising on architecture, fortifications, drains, water supply and anything else of a technical nature. His greatest achievement in Milan was the painting of *The Last Supper* (1495–98), considered by some to be the finest painting ever.

His most time-consuming project, though, was a giant bronze statue of Ludovico on horseback. This became the most celebrated of Leonardo's many over-ambitious failures, and when he met Michelangelo, the great sculptor mocked him for it. Antagonized, the two artists embarked on a duel by paint – both were to paint a giant mural of a battle scene. Leonardo began painting *The Battle of Anghiari*, and Michelangelo *The Battle of Casciari*. Neither artist finished their work. But it was in preparing for this painting that Leonardo did much of his anatomical research, spending time in Florence's Santa Maria Nuova hospital studying injuries and also dissecting corpses. He later planned to publish his anatomical manuscript in 1510, but the plan came to nothing.

Leonardo on the move

From 1500 on, the political instability of the times meant that Leonardo was constantly on the move between Florence, Venice, Rome and various other Italian cities, never staying more than a year or so in each. For a few years he was in the service of the ruthless Cesare Borgia, and he travelled across Cesare's lands, surveying them with many techniques that anticipated modern cartography. In 1503, Leonardo surveyed and planned a route for a canal to connect Florence to the sea. In 1505, at about the time he was losing interest in *The Battle of Anghiari* and painting the famous *Mona Lisa*, he wrote a book about the flight of birds, and over the next 2 years he filled his notebooks with ideas for flying machines, including a helicopter and a parachute.

By the time he was 60, Leonardo was beginning to tire of having to move all the time. He accepted a gift of rooms in the Vatican in Rome in 1513 and stayed there 3 years, hoping for a commission, but all the projects were going to others, such as Michelangelo and Bramante. Leonardo probably worked for a while as consultant on the Pope's plan to reclaim the Pontine Marshes, but there was not much else. He may also have spent his time investigating optical puzzles and experimenting with flight. So in 1516, the ageing artist accepted an invitation from King François I of France and left Italy for good.

Leonardo in France

In France, Leonardo took up residence in a house in Cloux, provided by François, and settled down to work on his notebooks. By this time he had had a stroke and his right hand was partially paralyzed, so work was slow. Fortunately, the King did not expect any substantial work from him. All Leonardo had to do was to produce plans for festivals and plays, and provide designs for toys, such as a mechanical lion that opened to reveal lilies in its breast. Indeed, the French king was a great admirer of Leonardo and was genuinely pleased to have him around. Poor Leonardo was constantly interrupted by visits from the well-meaning François, who walked through the tunnel that connected his Amboise palace to Leonardo's house.

Leonardo died quietly at Cloux on 23 April 1519 and was buried in the nearby church of St Florentin. (The church was destroyed in the French Revolution and Leonardo's remains were lost.) His young companion, Francesco Melzi, was grief stricken and stayed in the house for months before finally packing up all Leonardo's belongings, including 13,000 priceless pages of his notes, and heading off in a cart back to Vaprio in Italy. There the notes remained in his care until he died and left them to his son Orazio to look after.

Orazio had no interest in Leonardo's notes, stuffing some of the notebooks in a cupboard in the attic and giving others away. As word spread about the notebooks, so collectors descended on Vaprio, taking bits away and often ripping pages out, scattering Leonardo's work. Today the pages are distributed among museums and private collections, including the famous *Codex Leicester*, bought by Microsoft founder Bill Gates in 1994 for 30 million dollars. Of the original pages, almost half have been lost.

Leonardo's flying machines

Leonardo's notebooks reveal an astonishing range of inventions – clocks, printing presses, drills, boats, diving suits, cars and battle tanks. Yet perhaps the most astonishing are his flying machines.

At the heart of Leonardo's thinking was his belief that human and animal bodies are simply organic machines. This belief underpinned many of his investigations into anatomy, and also inspired many of his inventions. It was looking at birds flying that convinced Leonardo that a flying machine was feasible. 'A bird is simply an instrument functioning according to the laws of nature,' he wrote, 'in which case a man can recreate that instrument.'

His first designs for flying machines depended on flapping wings and are known as ornithopters. In 1487, he made a drawing of an ornithopter in which the pilot lay flat on a frame with his feet in stirrups, pedalling to make the wings flap. A few years later he came up with a design for an ornithopter with rudders and elevators to give control in flight – a remarkably advanced idea.

Ornithopters would never have worked because human muscle power simply isn't great enough. Leonardo may have realized this himself because he soon dropped the idea of flapping wings in order to work on gliders. To further his knowledge of wings and airflow, Leonardo studied birds and falling leaves, and in doing so invented the world's first anemometer to measure wind speed.

Ten years before he died, he drew a design for a glider that had a genuine control system not unlike modern hang gliders. Leonardo wrote, 'This [man] will move on the right side if he bends the right arm and extends the left; and he will then move from right to left by changing the position of the arms.' Recently, experts have built a machine based exactly on his design using only materials that would have been available to Leonardo – and proved that it would not only have been able to fly, but could also be controlled in flight, something not achieved until the Wright Brothers famous flight in 1903.

Leonardo realized that wings are not the only way of flying. He designed a helicopter to climb vertically into the air carrying men on a platform beneath it. Unlike modern helicopters, it did not have rotor blades, but a spiral screw designed to lift it up through the air. Helicopter toys with rotors had actually been around for centuries, but Leonardo was the first – and perhaps the last for another 500 years – to try and design one for lifting people.

Leonardo clearly grasped that the air has enough substance to support aerofoil shapes, and designed a machine based on a boat with oars in which the pilot rowed through the air. Although it clearly would not work, he had at least recognized the principle that was to lead to propellers and rotors centuries later.

Leonardo's anatomy

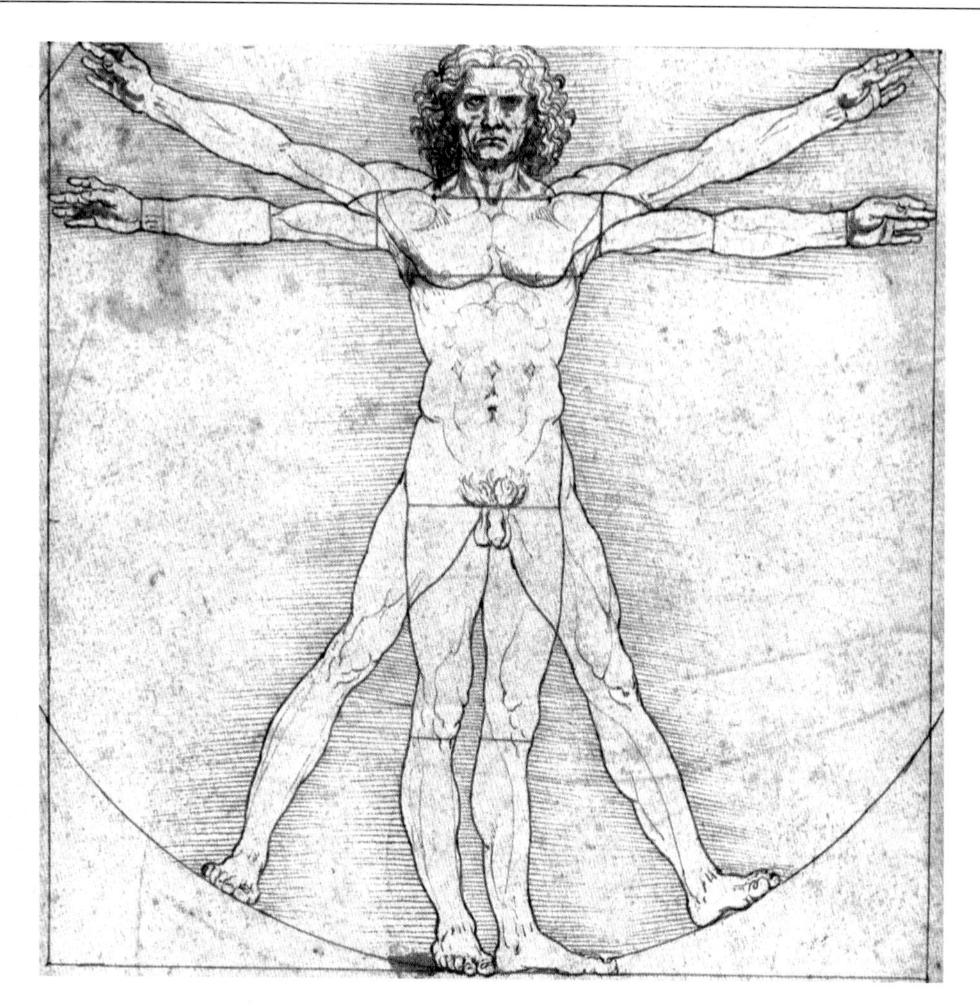

It is perhaps in his work on anatomy that Leonardo had the most lasting impact on the progress of science. At a time when most physicians were learning about the human body from the second-century physician Galen, Leonardo dissected corpses to find things out for himself. He was not the only one to do this, as it was expected that artists should know their anatomy. At around the same time as Leonardo was in Florence, for instance, a Florentine painter called Antonio Benivieni wrote a treatise based on his own dissections.

Yet Leonardo went much, much further in his analysis, dissecting over thirty human corpses personally, and conducting many experiments to see how parts of the body work. He also dissected bears, cows, frogs, monkeys and birds to compare their anatomy with that of humans. Leonardo's superlative skill in illustration and his obsession with accuracy made his anatomical drawings the finest the world had ever seen. In order to display the layers of the body, he developed the drawing technique of cross-sections that is still used today. He also developed remarkable three-dimensional arrays of muscles and organs from different perspectives, a technique that has only come into its own in the computer age.

One of Leonardo's special interests was the eye, and he was fascinated by how the eye and brain worked together. He was probably the first anatomist ever to see how the optic nerve leaves the back of the eye and connects to the brain. He was probably the first, too, to realize how nerves link the brain to muscles. There had been no such idea in Galen's anatomy.

It is perhaps in his study of muscles where Leonardo's blend of artistry and scientific analysis is best seen. He observed exactly what happened to muscles when they moved the body in different ways, how muscles in the face made people smile or frown, and much more.

Nicolas Copernicus

Nicolas Copernicus (1473–1543) was the sixteenth-century priest and astronomer whose ideas finally led to the realization that the Earth is not fixed at the centre of the universe, but is just one of the planets revolving around the Sun.

1473–1543

At the time Copernicus was born, Europe was beginning to feel the effects of the Renaissance. Classical ideas and texts were appearing in Europe from the Arab world, and thinkers there were beginning to add to them ideas of their own. At this time the Roman scholar Ptolemy's model of how the universe worked, described in his book *Almagest* (the *Greatest*), was still considered correct. In this model the Earth was still and fixed at the centre of the universe. Round the Earth were a series of invisible concentric spherical crystal shells in which the Sun, Moon, planets and stars revolved in perfect circles, one heavenly body in each, except for the stars.

Unfortunately for this model, actual observation showed that only the stars appeared to move in perfect circles. To account for this, Ptolemy suggested two main mechanisms – epicycles and equants. These explained the apparent motions of the planets while still keeping the idea of perfect circles. Epicycles were basically tiny circular motions inside each sphere, or wheels within wheels. Equants allowed the circles of the Moon and the planets to be slightly offset to turn around different points, called equant points, rather than all turning around the exact centre of the Earth.

This view of the universe as a series of nested crystal spheres turning round the Earth pretty much worked, in that it allowed astronomers to predict accurately the motions of the Sun, the Moon and the planets then known – Mercury, Venus, Mars, Jupiter and Saturn. But there were problems. In the 1490s, when Copernicus was in his twenties, the German astronomer Johannes Müller (better known by his Latin name Regiomontanus) published a summary of Ptolemy's *Almagest*, along with a critical commentary called the *Epitome*. In the *Epitome*, Regiomontanus pointed out that one of the problems with the Ptolemaic system is that if the circle of the Moon is offset as the system says it is, then it should get larger and smaller as it moves closer to the Earth and further away – and yet it clearly doesn't.

Another problem with the Ptolemaic system, as far as the young priest Copernicus was concerned, was that it seemed too elaborate and intricate. Surely God would have created something more simple and elegant? All these complications would disappear, Copernicus realized, if the Sun is at the centre and the Earth revolves around it, along with all the other planets. The only difficulty then was how to account for the fact that the Moon turned round the Earth. Although this 'heliocentric' (sun-centred) system had the great advantage of actually being true, it took more than a century for his idea to be widely known and even longer for it to be widely accepted (see box on p. 34).

Mikolaj Kopernik

Copernicus was born in Torun on the Vistula River in northern Poland on 19 February 1473. His real name was Mikolaj Kopernik, and it was only later in life that he adopted the Latin version of his name, Nicolas Copernicus. His father, a well-to-do merchant, died when he was about 10, and so he was brought up by his uncle Lucas Waczenrode, soon to be bishop of Varmia. His uncle saw to it that he had the general education typical for those destined for a career in the Church.

At the age of 20, Copernicus went to the University of Krakow to study the liberal arts, including astrology and astronomy. Then, 5 years later, he went on to study in Bologna in Italy, where he lodged for a while in the house of the distinguished astronomer and astrologer Ferrariensis. It was Ferrariensis who really inspired Copernicus's interest in the stars and introduced him to Regiomontanus's *Epitome*. In 1497, Copernicus observed an eclipse of the Moon in Bologna.

By the time Copernicus had finished his doctorate in canon law in 1503, he was well grounded in astronomy and already beginning to develop his ideas about a heliocentric universe. His uncle arranged for him to become canon at Frombork (Frauenberg) Cathedral in Poland, a post which allowed him plenty of free time both for studying astronomy and for pursuing various other tasks. He worked for the community as a doctor, for instance, and developed a plan for reforming the currency. In 1519 he was pressed into military service to command the defence of the castle of Allenstein against invading Teutonic Knights.

The great idea

In the meantime, he took advantage of his position in the cathedral to consolidate his ideas. Most of his astronomy was entirely on paper or in his head, but sometimes he climbed the cathedral tower at Frombork, and also at Allenstein and Heilsberg, to gaze at the night sky. Unlike later scientists, Copernicus had no interest in verifying his ideas by observation or experiment.

In 1514, Copernicus published a little handwritten book for his friends. Called *Commentariolus*, it gave the first outline of his revolutionary theory. In *Commentariolus* Copernicus included not just the idea that the Earth moved around the Sun, and that the stars are very, very far away, but the suggestion that this arrangement explains a number of phenomena, such as the retrograde motion of the planets. Ptolemy had explained retrograde motion – the fact that the planets appear to loop back on themselves every now and then – by means of complicated epicycles. But if it is accepted that the Earth is moving around the Sun with the planets, no such elaborate explanation is needed; retrograde motion is simply due to the changing view of the planets from the Earth.

Copernicus also suggested in *Commentariolus* that the time taken for each planet to complete its cycle through the night sky might increase the further it is from the Sun. Mercury's cycle takes 88 days, which makes it the nearest planet to the Sun. Venus takes 225 days, Earth 1 year, Mars 1.9 years, Jupiter 12 years and Saturn 30 years. It was thus easy for Copernicus to work out the order of the planets.

Keeping quiet

Copernicus made it clear in this little book that he was planning a major work to expound his theory in full, writing, 'Here, for the sake of brevity, I have thought it desirable to omit the mathematical demonstrations intended for my larger work.' This larger work, the famous *De Revolutionibus Orbium Coelestium* (On the revolutions of the heavenly spheres) was not published for a further 26 years, by which time Copernicus was on his deathbed. This delay was perhaps due to the religious view of the universe at the time: Copernicus may have thought it simply too dangerous to publicize his ideas. Others have suggested Copernicus delayed because he had not developed his ideas and proofs sufficiently.

The daring young man

Whatever the truth, the final spur Copernicus needed to finish his great book came when he acquired a disciple in the form of Georg von Lauchen, known as Rheticus, a young professor of mathematics from Wittenberg. Rheticus came to Frombork to learn more about Copernicus's ideas, and when he realized their importance he was determined that Copernicus should publish his work.

In 1540, Rheticus published a curtain-raiser. Called *Narratio Prima de Libtus Revolutionum Copernici* (First account of the revolutionary book of Copernicus), it summarized Copernicus's main idea that the Earth moves around the Sun. This seems to have done the trick: Rheticus wrote to a friend on 9 June 1541, he 'had finally overcome [Copernicus's] reluctance to release his volume for publication'.

The Planisphaerium Copernicanum – the heliocentric planetary system of Copernicus. Taken for granted in the modern era, this was a revolutionary view in Copernicus's day.

By August of that year, *De Revolutionibus Orbium Coelestium* was ready. Rheticus undertook to see it printed and took it to Johann Petreius, the best printer in Nuremberg. Unable to see the printing through personally, Rheticus deputed the task to a Lutheran minister called Osiander.

Although the story only emerged gradually, it seems that Osiander took it upon himself to write an unsigned preface saying that Copernicus's ideas were in no way intended to describe reality; they were simply a mathematical model to help with calculations. Osiander even changed the title to make it sound less definitive. Osiander was clearly worried about how people would react to Copernicus's revolutionary ideas. When Rheticus discovered this, he was livid, scrawling a huge red cross through the preface in his copy of the book.

What Copernicus thought about all this, no one quite knows, for he died of a stroke shortly after publication of the new book in 1543. It is said that he was handed the new book for the first time when he briefly recovered consciousness, and that he died with it in his hands: one cannot but hope the story is true.

Moving the Earth

It is not clear whether Osiander's preface to *De Revolutionibus* succeeded in limiting the outrage that might have been caused by Copernicus's ideas. The plain fact is that few people took much notice at first. Many of the original edition of 400 copies were left unsold, and there was certainly no clamour from the Catholic Church for Copernicus to be burned at the stake. In fact the only protest came from the Protestants Osiander had tried to appease in his preface. It seems likely that few people appreciated the real implications of Copernicus's ideas at first. Those who did understand them often remained quiet about it, and so no fuss was made.

One of those who did understand Copernicus was the English astronomer Thomas Digges, who wrote the first explanation of the Copernican system in English in 1576. Digges actually went further than Copernicus, suggesting that the universe around the solar system is infinite, with a multitude of stars in all directions.

The famous Danish astronomer Tycho Brahe (1546–1601) did not accept the Copernican model himself, yet his meticulous observations slowly built up a pile of evidence in favour of it – not least of which was the observation of a new star, a supernova, in 1572 which showed that the stars were not perfect and unchanging after all, whatever the Church and Ptolemy said.

One of the reasons why Brahe could not accept Copernicus's model was because he was a very precise observer – and Copernicus's model did not precisely fit the facts. This is where Brahe's assistant Johann Kepler (1571–1630) comes in. Unlike Brahe, Kepler did accept the Copernican model, and what is more, in a brilliant feat of mathematical inspiration, he found a way to make it fit the facts, using Brahe's observations. If the orbits of the planets and the Earth are elliptical, not circular, Kepler realized, then the Copernican system fits the facts perfectly.

Kepler's ideas were finally published in full in 1619 in his book *Harmonice Mundi* (Harmony of the world), but by that time another astronomer, Giordano Bruno, had been burned at the stake. Because Bruno was a Copernican, and because like Digges he believed in a universe filled with infinite stars, people assume he was burned for his dangerous astronomical ideas. In fact, he was condemned by the Inquisition for his 'blasphemous' Arian beliefs and his practice of magic.

All the same the Catholic Church began to associate the Copernicanism with the widespread threat of the Protestant Reformation. In 1610, Galileo saw through his telescope the clinching evidence that Copernicus was right – moons circling Jupiter and moonlike phases for Venus. As Galileo began to publicize his ideas – right from the heart of Catholic Europe, in Florence – the Church finally decided to take action.

In 1616, 73 years after its first publication, they banned *De Revolutionibus*. That same year, the cardinals summoned Galileo to Rome and forbade him to talk about Copernicanism. Galileo persisted, and in the end the Cardinals had to threaten him with torture to shut him up. But of course the battle was lost. The Copernican revolution was in full swing all over Europe. However, it would be another 200 years before the Catholic Church lifted its ban on *De Revolutionibus*.

Andreas Vesalius

Nothing is closer to us than our bodies, yet it took as long to explore the human body as it did the Earth. Andreas Vesalius (1514–64) was perhaps the greatest body explorer of all, and his book De humani corporis fabrica *was the first great landmark in the discovery of human anatomy.*

1514–64

Until the time of Vesalius in the early sixteenth century, knowledge of human anatomy was based essentially on blind faith and guesswork. Direct observation played very little part at all. Remarkably, medical students of Vesalius's time learned human anatomy not by studying bodies, but by reading the works of the Roman physician Galen (AD129–c. 216). Students went to lectures where corpses were dissected by a barber (basically a man with sharp knife) – even this was an innovation – but never themselves dissected, preferring to read from Galen's works while the barber sliced away in the distance.

There was no doubt that Galen had been a great authority, and in his time he was a wonderful physician, perhaps the most skilled in the ancient world, but his real knowledge of anatomy was shaky at best. Yet such was his influence that his work was never questioned until Vesalius's day.

One of the problems was the taboo on dissecting human bodies. A fourteenth-century Bolognese doctor called Mondino DeLuzzi was one of the first to carry out a dissection, yet such was the hold of Galen's works on him that De Luzzi generally failed to see what must have been staring him in the face, and simply compounded Galen's errors.

What was remarkable about Vesalius is that he broke two taboos. First of all, he dared to challenge the 1,300-year-old authority of Galen. Secondly, he did so by actually dissecting human bodies, looking at real anatomy closely and encouraging his students to do the same. All the anatomical knowledge he gained in this way was put into his great book *Fabrica*, which laid the foundations of modern medicine.

The young Vesalius

He was born into a Flemish family in Brussels in 1514. His father had been a court official in the service of the Holy Roman Emperor, and so had his grandfather, and it was his ambition to emulate them. There is no doubt that Andreas was an extremely driven and determined young man. As a teenager, he began medical studies at Louvain University in Brussels, and his fanatical desire to study anatomy was already in evidence. At 16, he was already stealing corpses for dissection from gibbets in the middle of the night, and he soon began to ask judges to set the dates of executions to days suitable for his work.

Before long, his pursuit of corpses was becoming positively gruesome. In 1533, he went to study in Paris, where he was often out in the night scouring graveyards for fresh bodies, or scavenging the poorer burial sites, where he would fight with wild dogs over a body. Extraordinarily, to avoid unwanted attention, he took the bodies back to his bedroom and secretly dissected them there. Night after night he would sleep with a rotting, partially dissected corpse beside him. Since he kept corpses for several weeks, the stench must have been truly appalling.

But his obsessive pursuit of anatomical knowledge was paying off. While at Paris, his skill came to the attention of Jacob Sylvius and John Guinter, the two greatest anatomists in Europe, both of whom were teaching at the university. Just a year later and aged only 23, he was made head of the Department of Surgery and Anatomy at the University of Padua in Italy, then the most prestigious medical school in the world.

Vesalius in Padua

Vesalius, like all doctors of the time, was still reading and learning from Galen, but he abandoned the age-old practice of reading to students while a barber dissected a corpse, opting instead to perform the dissection himself while describing to his students exactly what he was uncovering. Vesalius insisted that to know the human body you must dissect it. This in itself was revolutionary enough, but, of course, Vesalius was also beginning to discover that Galen was not always right.

In 1538, Vesalius got John Stephanus of Calcar, an artist in Titian's studio, to draw versions of six of the charts he had been sketching for his students. Published as *Tabulae anatomicae sex* (Six anatomical charts), three of these charts showed views of the human skeleton, while the other three showed the portal vein near the heart, the heart and all the body's veins, and the heart and all the arteries. This was a tremendous novelty. Few anatomical works before had ever been illustrated, and Vesalius's mentor Sylvius was annoyed, feeling that illustrations would mislead students and degrade scholarship. Worse still, Vesalius's charts corrected some small but important errors of Galen.

Despite the opposition it provoked, the *Tabulae* was an instant hit with students, and Vesalius's reputation as an anatomist began to grow, as did his knowledge of human anatomy. Fired by the success of the *Tabulae*, he embarked on a huge and groundbreaking project – to create the first comprehensive, accurate, illustrated guide to human anatomy based on dissections.

Working with a brilliant team of artists, which may have included John Stephanus, Vesalius laboured for four years to produce this masterwork. When everything was finally prepared, he sent the manuscript across the Alps to Basel to John Oporinus, a distinguished professor and wonderful printer, who was to print the book on the finest paper with the best typography. So concerned was Vesalius that his book should be of the best quality and highly accurate that he himself rode over the mountains to Basel and stayed there to oversee the printing personally.

In late summer 1543, when Vesalius was still just 29, *De humani corporis fabrica* was finished, and he sent a magnificent purple silk-bound presentation copy to the Emperor Charles V, complete with over 200 fabulous hand-coloured illustrations. The Emperor was so impressed that a few months later, Vesalius was invited to become one of Charles's personal physicians.

The royal physician

Within less than a year, Vesalius's burning thirst for knowledge seems to have evaporated, and he abandoned his academic career entirely. Having achieved his ambition to become a court official, he settled down to a distinguished but conservative career, marrying a Brussels girl, Anne van Hamme. The couple had a child the next year, who was also named Anne.

Vesalius later implied that his decision to abandon research and academia was partly due to the stream of vicious criticism he had received for publishing *Fabrica* – criticism that he said 'gnawed at my soul'. Rather than face such attacks again, he would remain at court 'far from the sweet leisure of studies' and he 'would not consider publishing anything new even if I wanted to do so very much'.

Yet though he gave up research, there was no doubt that he became a highly distinguished physician, esteemed across Europe. When Henry II of France was injured in the head by a lance in a joust in 1559, Vesalius was summoned to Paris to help the court physicians, who were trying all kinds of desperate measures to save the king. When Vesalius arrived, he examined Henry and was instantly certain that nothing could be done for him. Risking the French court's wrath, he announced that the king would soon die – and ten days later he did, despite all the efforts of Europe's best doctors. Vesalius was at the autopsy which revealed that the lance blow had inflicted massive damage to Henry's brain.

Three years later, Don Carlos, the Crown Prince of Spain, smashed his head tumbling downstairs in his hurry to catch a glimpse of the caretaker's beautiful

The books of Fabrica

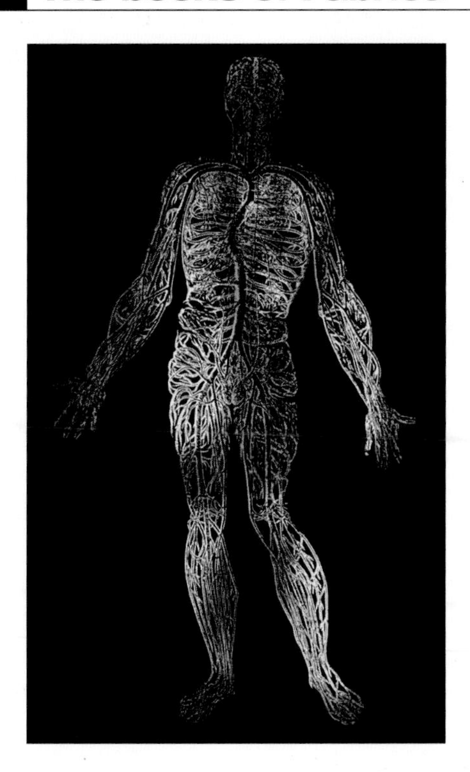

The full title of Vesalius's book was *De humani corporis fabrica, libri septem*, for it was actually constructed in seven volumes (*libri septem*).

Book 1 reveals Vesalius's understanding of the importance of the skeleton. Before Vesalius, no one had grasped just how important bones are in terms of giving the body its shape and movement. However, he gave over the whole of Book 1 to bones, all drawn with wonderful detail and verve. The book ends with three full-page drawings of the complete skeleton adopting different poses, including one suspended from a gibbet and another leaning on a desk studying anatomy!

Book 2 is about muscles, and the drawings in this volume are as superb as those in Book 1. Books 3 to 7 are less spectacularly beautiful, but still very finely illustrated. Book 3 is about the veins and arteries, Book 4 the nervous system, Book 5 the main body organs, Book 6 the heart and lungs, and Book 7 the brain.

At the same time as *Fabrica* was published, so was a compact single volume version called *Epitome*. *Epitome* contained many of *Fabrica*'s fabulous drawings, but it was designed to be carried to the dissecting table by medical students – reinforcing the message that Vesalius had an intensely practical, dissection-based approach.

A false-colour version of an engraving of the venous system from De Humani Corporis Fabrica *(On the Structure of the Human Body).*

daughter. Again, as the young prince's condition deteriorated, Vesalius was summoned for his expert advice. Arriving in Madrid, Vesalius took charge of the treatment. At first, Don Carlos seemed to grow even more ill, but Vesalius's treatment began to work, and after a few months he was fully recovered. King Philip II was convinced that the cure came from the mummified corpse of Fra Diego, a Franciscan friar of the thirteenth century, which had been placed next to the unconscious Don Carlos. But Vesalius's reputation could not have been higher.

In 1564, Vesalius set out on a pilgrimage to the Holy Land. The reason for this is not known. One theory is that, when dissecting a corpse of a young aristocrat in his Paduan days, he had been horrified to find his subject breathing, and the dissection had gone too far for Vesalius to save him. The tragedy had long haunted Vesalius, and the pilgrimage was perhaps in penance for this ghastly mistake. Whatever the truth, Vesalius never returned, dying on the ship on his way home from Jerusalem.

Vesalius's *Fabrica*

Vesalius's *Fabrica* was without doubt the greatest medical book ever produced until that time. It was impressive in size alone at 42 cm tall and 28 cm wide, and it had over 700 densely packed pages. It was also a beautiful book. Its 200 illustrations were not merely the first complete collection of accurate anatomical drawings ever done, but were also artistic creations of the highest order. The skeletons were not just flat representations of bones, but imaginatively posed and presented pictures.

But it was neither the beauty nor quality of the book that were its most remarkable aspects. What made *Fabrica* a milestone in scientific history was the merciless, stark accuracy of its representation of the human body, flayed open in dissection so that not the least corner was left hidden. Never before had the workings of the human body been represented so precisely. It was by no means flawless in its accuracy, and three of Vesalius's protegés, Gabriel Fallopio, Bartolomeo Eustacius and Realdo Colombo, very quickly made major new anatomical discoveries. But it was right on all the major details – a remarkable achievement after thousands of years of vagueness.

Just how innovative *Fabrica* was – in particular in its emphasis on dissection – can be gauged from the savagery of some of the criticism it received. Remarkably, some of the most stinging comments came from Jacob Sylvius, Vesalius's mentor at Paris. In an open letter to the Emperor Charles V, Sylvius wrote: 'I implore His Imperial Majesty to punish severely, as he deserves, this monster born and bred in his own house, this worst example of ignorance, ingratitude, arrogance, and impiety, to suppress him so that he may not poison the rest of Europe with his pestilential breath.'

But despite such criticisms, *Fabrica* was an immediate and huge success. Within a decade, any medical student worth his salt would be learning anatomy not just from Vesalius's book, but by dissecting corpses for himself.

Inspired by Vesalius's work, other physicians began to make their own dissections, gradually filling the gaps in our knowledge of human anatomy. They began to learn about physiology, too, and within a century understanding of the human body and how it works had advanced to an extraordinary degree – and the foundations were laid for the remarkably detailed picture we take for granted today.

Galileo Galilei

Galileo Galilei (1564–1642) was the first great scientist of the modern age. His insistence on observation and experiment laid the foundations for the scientific revolution of the seventeenth century. The force of his ideas, however, would bring him into head-on conflict with the Roman Catholic Church.

1564–1642

A BRILLIANTLY CREATIVE MAN, GALILEO ACHIEVED MANY SCIENTIFIC FIRSTS, each of which would have been enough to give him a place in history. For one thing, he was an ingenious inventor, and among the most notable of his ideas was the value of the pendulum as a timekeeper, which led to the creation of the first accurate clocks. Another Galileo invention was the thermometer. He invented the sector, too, the first simple device for calculating the trajectory of a missile. And he could be said to have invented the astronomical telescope.

Above all, though, he was a great scientist. For example, he did not simply take the telescope and turn it into a major scientific instrument. He had the insight to use it to look at the night sky and make many extraordinary discoveries there – including the mountains and valleys on the Moon's surface; the moons of Jupiter; the fact that Venus has phases like the Moon; and the Sun has spots. It was these discoveries that persuaded Galileo that Copernicus's view that the Sun and not the Earth was at the centre of the universe was true. This led to his clash with the Catholic Church, which insisted on the old Ptolemaic view of the universe, in which the Earth is fixed immobile at its centre.

But perhaps Galileo's greatest achievements were in understanding how things move, which created the basis for the modern science of physics. For almost 2,000 years, people had accepted the views of Aristotle on how things fall, why things stop and go, and how things get faster and slower – and remained blind to the evidence of their senses. Galileo overturned Aristotle's apparently common-sense views – and paved the way for Newton's full understanding of force, motion and gravity half a century later.

It was Galileo's insistence on the importance of demonstration, observation and experiment that proved Aristotle wrong, and led to insights and proofs of entirely new ideas. Galileo was not alone at the time in looking at things this way. The English thinker Francis Bacon was a pioneer of these new methods. However, Galileo put them into practice with such force and insight, and with such crucial effect, that he deserves to be called, as he often is, 'the father of modern science'.

The young Galileo

Galileo Galilei was born in Pisa in Italy on 15 February 1564. His father Vincenzo was descended from a Florentine family that had fallen on hard times. Vincenzo was a musician of some talent and a highly independent, combative turn of mind, which the flame-haired Galileo inherited.

At the age of 10, the young Galileo was sent to school at the monastery of Vallombrosa. He took so well to the monastic life that at 14 his father took him away from the school, worried that his son might take up the life of a poor man of the cloth. He spent a few years with tutors in Florence, and then his father brought him back home to study medicine at Pisa University.

The young Galileo began to challenge his teachers. He would stand up and question lecturers on the absurd fixity of their ideas, which came largely from Aristotle. Why, Galileo would ask, do hailstones all hit the ground at the same speed if heavier things fall faster, as Aristotle said? And then he would laugh when the lecturer tried to suggest it might be because heavier stones come from higher up. Clearly, as Vincenzo realized, Galileo would never possess the calm bedside manner of a doctor.

Then Galileo discovered mathematics – and the works of the Greek geometer Euclid (see p. 8). Euclid insisted on clear proofs and demonstrations before anything could be accepted as true – and this idea stuck with Galileo for the rest of his life. He was also struck by how that other great Greek mathematician, Archimedes, had begun to apply this approach to science as well as mathematics. 'Those who read his works,' Galileo wrote, 'realize only too clearly how inferior all other minds are.'

Discovering mathematics

Vincenzo arranged for Galileo to be tutored in mathematics by the brilliant Florentine court mathematician Ostilio Ricci, and Galileo quickly outshone his teacher. Even at this young age, Galileo was highly inventive. According to one story, he was sitting through a dull sermon in Pisa Cathedral one Sunday, idly watching a lamp swinging on a long wire. He suddenly noticed that no matter how wide the lamp swung, it always completed its swing in exactly the same time. Proving this with a series of simple experiments at home, he realized that it could be used to make a timing device, which he called a *pulsilogium*, because it could be used to time a patient's pulse. This later became the basis for the pendulum clock.

At the age of 21, Galileo began to teach mathematics, and this was to be his main source of income for the remainder of his life. At first he tutored privately. Later he was appointed lecturer at Pisa University, but the pay was poor. When his father died in 1591 and Galileo was left with the whole family to support, he moved to a better-paid post at Padua and remained there for 18 years. He was a distinctive figure who strode around the precincts in dishevelled clothes and without the regulation academic gown. It was in Padua that Galileo took up with Marina Gamba, a fiery back-street beauty who was described in the slang of the day as *una donna di facile costume* (a girl whose clothes come off easily). They never actually lived together, or married, but they had three children and were effectively a family.

In between his teaching work, Galileo began to investigate scientific problems that took his interest. His first ideas on how things move were encapsulated in 1590 in a series of essays called *De motu*, written while he was at Pisa. It was while working on this that he conducted his famous experiment on the Leaning Tower of Pisa. To show the error of Aristotle's notion that heavier things fall faster, he dropped cannonballs of different sizes and weights from the tower's overhang and showed beyond doubt that they land at the same time. These ideas were later developed in *La meccaniche*, which combined mathematics and physics to create the new science of mechanics – the science of force and motion.

Galileo's telescope

Sometime in the summer of 1609, Galileo visited Venice and became intrigued by a novelty called a perspicillium, made by a Dutch spectacle maker. It consisted of two lenses in a tube and could make a distant steeple look as if it was just across the way. Inspired, Galileo realized how it worked and immediately made one of his own with ten times as much magnification. He called it a telescope and it rapidly became famous throughout Italy.

Galileo's trial for heresy by the Catholic Church. Threatened with death, Galileo finally renounced his views on 22 June 1633. He was imprisoned in his villa at Arcetri. In 1642 the Church refused to bury his body on consecrated ground. His conviction was finally annulled in 1992.

In a stroke of genius, Galileo used his telescope at night to look at the Moon and the stars. At once he saw the Moon was not the perfect sphere it was supposed at the time to be, but had mountains, valleys, cliffs and maybe even seas. Soon he discovered that Jupiter was not the perfect isolated sphere it was supposed to be either, but actually had four moons of its own. And then he noticed that Venus went through phases, just like the Moon, as our view of it from Earth changes. In 1610, Galileo published all these discoveries in a stylish work in Latin called *The Starry Messenger*.

Galileo against the Church

Galileo's discoveries clearly implied that the Earth was not at the centre of the universe, as most people believed then, but moved round the Sun, as Copernicus had suggested 70 years earlier. However, there was no mention of this in *The Starry Messenger*. By the time he moved to Florence to become 'philosopher and mathematician' at the court of Grand Duke Cosimo de Medici, Galileo was certainly convinced that Copernicus was right, and was saying things like, 'Earthly laws apply to the heavens'.

Rival academics were still firmly wedded to Ptolemy's model of the universe, which put the Earth immobile at the centre, with the Sun, Moon, planets and stars revolving in perfect, unblemished layers around it. These ideas fitted in neatly with

Galileo and motion

The philosophers of ancient Greece had known a great deal about statics – that is, about things that are not moving. But they were often lost when it came to questions of dynamics, or how things move. They could see, for instance, that a cart moves because the horse pulls it. They could see, too, that an arrow flies because of the power of the bow. But they had no explanation for why an arrow goes on flying through the air when there is nothing to pull it like the horse pulls the cart. The famous Greek philosopher Aristotle made the common-sense assumption that there must be a force to keep something moving – just as a bike will only keep on moving if the rider pushes on the pedals.

But common sense can be wrong, and it was Galileo's genius to see the superiority of practical observation and experiment, or *cimento* as he called them, over common sense. After a series of *cimento* – many involving rolling balls down slopes – Galileo realized that force was not needed to keep something moving. Exactly the opposite is true: something will keep moving at the same speed unless a force slows it down. This is why the arrow goes on flying through the air. It only falls to the ground because the resistance of the air (a force) slows it down enough for it to be pulled to the ground by gravity (another force). This is the principle of inertia.

Galileo resisted the notion of gravity because he felt the idea of what seemed to be a mystical force was unconvincing, but he was the first to appreciate the concept of inertia and verify it experimentally. He realized that there is no real difference between something that is moving at a steady speed and something that is not moving at all – both are unaffected by forces. But to make the object go faster or slower, or begin to move, a force is needed.

Further experiments, this time with swinging weights, led Galileo to a second crucial insight. In these experiments, he began to appreciate the notion of acceleration, and just what causes it. If something moves faster, he realized, then the rate at which it accelerates depends on the strength of the force that is moving it faster, and how heavy the object is. A large force accelerates a light object rapidly, while a small force accelerates a heavy object slowly.

These insights of Galileo's were very similar to the first 2 of the 3 laws of motion that Newton described 46 years later in his *Principia* (see p. 58). It is even possible that Galileo was aware of the third law – about action and reaction. Although he did not formulate his ideas with the same grand clarity and mathematical certainty of Newton, he did lay the foundations of our modern understanding of how things move.

Reproduction of Galileo's telescope, with the cathedral of Florence in the background.

the Bible, leaving the realm of the heavens purely in God's control. Galileo's beliefs could be seen as heretical, and he was denounced to the Inquisition as a blasphemer.

These were highly dangerous times for heretics. It was barely 30 years since the astronomer Giordano Bruno had been burned at the stake for his ideas. So Galileo went to Rome in 1616 to plead his case. His pleas fell on deaf ears. Copernicus's book (see p. 34) was banned at once and Galileo was sent back to Florence with a stern warning not to 'hold or defend' Copernican ideas. However, when the apparently more sympathetic Urban VIII became pope in 1624, Galileo went back to Rome to put his case again. Urban said that Galileo might write about 'systems of the world' as long as he did not defend Copernicanism.

At once, Galileo – by now over 60 years old – began writing a book in the form of a dialogue between three characters: the clever Sagredo (who argues for Copernicus), the dullard Simplicio (who argues hopelessly for Aristotle) and Salviati (who takes an apparently neutral line but is clearly for Sagredo). Called *The Dialogue*, the book was an instant success across Europe. Then the Jesuits pointed out to the pope that the ridiculous Simplicio might be based on him. At once Galileo was dragged back to Rome by the furious pope. There he was quickly forced by the papal authorities to deny – maybe under threat of torture – that the Earth moves round the Sun, and was sent away to be imprisoned in his own house for the rest of his life. Legend has it that as he was led away, he muttered *'eppir si muove'* (yet it does move).

Despite his age and ailing health, Galileo continued to do scientific research. In 1637, just before he went completely blind, he saw through his telescope that the moon wobbles on its axis. He eventually died on 8 January 1642, the same year that Newton was born in England. It took the Vatican 350 years to admit that 'errors might have been made' in the case of Galileo, but they need not have bothered. Even as he was carried to his grave, the scientific revolution begun by the Italian genius was getting into its stride, despite the pope's opposition.

'They were seen to fall evenly': the original caption to an illustration of Galileo's famous – if probably apocryphal – experiment on the velocity of falling bodies from the Leaning Tower of Pisa.

Christiaan Huygens

One of the few scientists of the late seventeenth century with a stature approaching that of Newton, Christiaan Huygens (1629–93) was famed for the invention of the pendulum clock, the first accurate clock, and such astronomical discoveries as Saturn's moon Titan.

Huygens lived at a time when science was just beginning to make its mark in the world, and the greatest minds in Europe were caught up in the fervour of scientific discovery. This was the time when Newton discovered gravity and the basic rules that governed every movement in the universe; the time when Newton and Liebnitz created the mathematics of calculus; when Hooke and Leeuwenhoek were discovering the world of microscopic life. Huygens was at the forefront of this scientific revolution, and lived at its hub in northwest Europe.

When Huygens was a growing boy in the Netherlands, Galileo was being threatened with torture by the papal authorities in Rome for suggesting that the Earth might not be the fixed centre of the universe. Yet not long after, Huygens was writing, 'We shall be less apt to admire what this World calls great ... when we know that there are a multitude of such Earths inhabited and adorn'd as well as our own.' Such sentiments would not sound out of place coming from the mouth of an astronomer today.

Like so many scientists of the day, Huygens's interests were wide ranging. Like Hooke and Leeuwenhoek, he made his own microscopes and made important discoveries in the microscopic world. Like Hooke and Newton, he made his own astronomical telescopes, which were technically superior to anything that had gone before. With them he discovered Saturn's moon Titan and the nature of Saturn's rings, mapped the surface of Mars for the first time, identified stars within the Orion nebula and discovered a number of other nebulae.

Above all, though, Huygens is famous for two crucial ideas. First of all, he invented the pendulum clock (see box on p. 48), the world's first accurate timekeeper and explored all the mathematics associated with pendulums – which led him and Hooke to an early prediction of the link between the elliptical orbits of the planets and the inverse square law of gravity (see p. 54).

Huygens is also famous for the proposition of the wave theory of light known as the 'Huygens Construction', which he outlined in his *Treatise on Light* (1690). This gave a much better explanation than Newton's rival particle theory (see p. 62) for the way light is reflected and refracted, and predicted, correctly, that light must travel more slowly in denser materials. Newton's status ensured Huygens's wave theory was neglected until the nineteenth century, when the Englishman Thomas

Young conducted experiments that seemed, after all, to prove the wavelike nature of light. Now scientists accept that light can behave as both particle and wave, but Huygens has been well and truly vindicated.

My little Archimedes

Born on 14 April 1629 in a grand house in The Hague, Huygens grew up in a sophisticated environment. His father Constantin was a diplomat of the Dutch Republic as well as a poet, composer and patron of the arts. There was always a string of distinguished visitors to the house on the Plein, including the English poet John Donne, the painter Rembrandt and, most significantly, the great philosoper and mathematician René Descartes.

The title page of Treatise On Light, *Huygens's classic work on physical optics, published in 1690.*

TRAITE
DE LA LVMIERE.
Où font expliquées
Les caufes de ce qui luy arrive
Dans la REFLEXION, & dans la REFRACTION.
Et particulierement
Dans l'etrange REFRACTION
DV CRISTAL D'ISLANDE.
Par Monfieur CHRISTIAN HUYGENS, Seigneur de Zeelhem,
Avec un Difcours de la Caufe
DE LA PESANTEVR.

A LEIDE,
Chez PIERRE VANDER AA, Marchand Libraire.
MDCXC.

Surrounded by such a galaxy of talent, the young Christian had plenty of role models, and he developed a lifelong love of learning that spanned languages, art, music, law, engineering and, above all, mathematics. His father called him 'mon Archimede'. But it clearly wasn't all books, for he became an excellent card player, a dab hand at billiards and a top-class rider.

At the age of 16, the young Huygens went to the University of Leiden, where he began to study mathematics and law in earnest. After two years at Leiden, he continued his studies at the University of Breda, and he was already beginning to make a name for himself through his correspondence with the great French mathematician Marin Mersenne, and later with his own elegant treatises on major mathematical problems.

Stargazing

Soon, however, Huygens's attentions turned to the skies. He began to develop better ways of grinding and polishing lenses for telescopes and it was not long before he was building some of the most powerful lens telescopes of the day. Using one of these telescopes, he discovered Saturn's moon Titan in March 1655. That year, he made his first trip to Paris, where he met some of the most distinguished thinkers of the day.

Back in Holland the following year, Huygens was able to see the true nature of Saturn's rings – that they are indeed a narrow band of rings around the planet and appear to change shape simply because we see them from Earth at different angles. Astronomers with inferior telescopes would not accept his description at first, but as telescopes improved Huygens was seen to be right. Working in astronomy demanded accurate timekeeping, and that year Huygens developed the pendulum clock (see box on p. 48), as described in his famous book *Horologium* (1658).

Society man

By now, Huygens's reputation was beginning to spread, and when he returned to Paris in 1660, he was something of a celebrity. He met the great French mathematician Blaise Pascal, and became part of a circle of the best thinkers of the day, including Pascal, Carcavi and Sorbiere. Huygens wrote to his brother that there was 'a meeting every Tuesday where twenty or thirty illustrious men are found together. I never fail to go ...', and in 1661 he went to England to find out more about the newly founded Royal Society. While in London, he showed the English scientists his telescopes, and they were suitably impressed. Huygens, in turn, was impressed with Hooke's air pump and Boyle's experiments with it, and soon began his own air pump experiments.

Two years later, Huygens was invited to join the Royal Society. Then in 1666 Huygens was invited to become one of the founding members of the French equivalent of the Royal Society, the Académie Royale des Sciences. He moved to Paris, very soon becoming the leading figure in the Académie, and by 1672 was its president, a unique honour for a Dutchman.

'We shall be less apt to admire what this World calls great ... when we know that there are a multitude of such Earths inhabited and adorn'd as well as our own.'

Huygens remained in Paris for 15 years, and it proved to be one of the most productive periods of his life. It was during this time that he developed his theories on light, and also further developed his pendulum clock to try and create a timepiece accurate enough to use on ships and solve the problem of determining longitude (see box on p. 48). He developed a magic lantern, an ancestor of modern film projectors, and an engine powered by gunpowder. He is also said to have taught mathematics to the later famous German scientist Leibniz.

Ill health

All the time, though, Huygens was constantly plagued by illness. In 1670, he had to return briefly to Holland, believing himself close to death. While lying ill in bed, he called for the English ambassador to give him his papers on mechanics, saying the Royal Society was 'an assembly of the choicest wits in Christendom' and that he would sooner deposit his 'little labours' in their hands than anywhere else. Huygens survived this time and returned to Paris, but his health became increasingly frail.

By 1681, Huygens was so ill that he was forced to give up his position in the French Académie and return to the Hague. He made a return visit to London in 1689, and met Newton there. Huygens and Newton both admired one another. Newton described Huygens as 'the most elegant mathematician of the age'. But on many points they did not agree. Newton thought light was made up of particles, while Huygens thought it moved in waves. Even more crucially, Huygens was not convinced by Newton's theory of universal gravitation: 'I esteem his understanding and subtlety highly, but I consider they have been put to ill use in the greater part of this work, where the author studies things of little use or when he builds on the improbable principle of attraction.' Like Galileo, Huygens thought the idea of an invisible force was just too fanciful.

The final frontier

Huygens returned to isolation in Holland again, and spent much of his time contemplating the nature of the universe. His ideas culminated in his book *Cosmotheoros*, which was translated from Latin into English in 1698 as *The Celestial Worlds Discover'd: or, Conjectures Concerning the Inhabitants, Plants and Productions of the Worlds in the Planets.*

This extraordinary book was the first serious scientific work on the idea of a

Huygens and time

Galileo first alerted the world to the timekeeping possibilities of a swinging pendulum. But it was Huygens who successfully tied it with an escapement mechanism – the weight and gear that kept the clock hands moving – to create the first pendulum clock, as he describes in his *Horologium* (1658).

Clocks using falling weights and gears were already in wide use, but they were wildly inaccurate, losing or gaining at least 15 minutes every day. Huygens's pendulum clocks could be accurate to within a minute or so in a week. Huygens was not slow to realize the commercial potential, and within a year had arranged for such clocks to be made under licence. In just a few years, the pendulum clock was being used for timekeeping across northern Europe. It was a crucial invention not only for the world in general, but also for the progress of science.

But Huygens himself was not satisfied. He wanted to make a clock that kept perfect enough time for it to be carried aboard ships and used for making longitude measurements. Any gain or loss in time could lead to huge miscalculations in position. Very large pendulum clocks in scientific establishments could be just about accurate enough, but it was thoroughy impractical to take such a clock on a ship. So Huygens's set about working out why smaller pendulum clocks were inaccurate.

Huygens discovered that a simple pendulum is not actually a perfect timekeeper after all. In fact, it completes smaller swings faster than big swings – so it is not 'isochronous' (keeping equal time). Any variation in the size of the swing would make a clock gain or lose time. Huygens realized that this was because the weight or 'bob' of the pendulum followed a circular path. Brilliantly, he showed mathematically in his great book *Horologium oscillatorium* (1673) that if the bob's path was a cycloid instead of a circle, it would be isochronous no matter what the length of the swing. (A cycloid is the curved path traced out by a point on the rim of a wheel as it rolls along.)

Huygens's inventive genius enabled him to go further and make the pendulum's swing cycloidal by suspending a rigid pendulum rod on two cords whose swing either way was limited by two plates called 'cycloidal checks'. Despite extensive sea trials, however, Huygens's cycloidal pendulum clocks never quite worked in practice, and it was another 100 years before English clockmaker, John Harrison, made a sufficiently accurate and robust 'chronometer' to use at sea and solve the longitude problem. But if the clock itself was not a total successs, Huygens's theoretical work was a milestone in science, playing a key part in the understanding of centrifugal force. It also laid the groundwork for Newton's laws of motion by showing how an object will travel in a straight line, unless pulled into a curved path by some other force.

In 1675, Huygens introduced another great innovation in timekeeping, the spiral balance spring. Just as a swinging pendulum regulates the motion of pendulum clock, so the spring balance regulates the motion of the balance wheel in a watch. The balance wheel is the finely balanced wheel that rotates to keep a watch mechanism operating. The spiral balance spring controlled its rotation so well that watches could keep time to within a day per year. Within a few years watches were being carried in gentlemen's pockets across Europe.

Computer artwork of the Cassini spacecraft orbiting Saturn (right). Titan, Saturn's largest moon, is at upper left. Cassini is carrying the disc-shaped Huygens probe, named for the discoverer of Titan (centre) on its underside.

gigantic universe populated with 'so many Suns, so many Earths'. 'How must our Wonder and Admiration be increased,' he wrote, 'when we consider the prodigious Distance and Multitude of the Stars.'

Even more startling were Huygens's discussions of the possibility of extraterrestrial life:

> *[A] Man that is of Copernicus's Opinion that this Earth of ours is a planet, carry'd round and enlighten'd by the Sun like the rest of them, cannot but sometimes have a fancy that it's not improbable that the rest of the Planets have their Dress and Furniture – nay and their Inhabitants too as well as this Earth of ours, especially if he considers the later Discoveries made since Copernicus's time of the [Moons] of Jupiter and Saturn.*

The book was much admired at the time, but it has taken three centuries for scientists to get to a position where they can seriously investigate the possibility of extraterrestrial life.

Cosmotheoros was Huygens's last work, and it was published posthumously. The 1690s, for Huygens, had been marked by rapidly deteriorating health and depression. He died after considerable suffering in March 1693.

Anton von Leeuwenhoek

Anton von Leeuwenhoek (1632–1723) was one of the most unassuming scientists who ever lived, spending all of his long life in Delft, where he was a draper. But in his back room he discovered an entirely new world – the world of microscopic life, including bacteria and protozoa, nematodes and rotifers, and human sperm and blood cells.

1632–1723

UNTIL LEEUWENHOEK STARTED TO WORK WITH HIS MICROSCOPE, no one suspected that there was any kind of life too small to see with the naked eye. Fleas were thought to be the smallest possible form of life, and they were clearly visible to the sharp eyed. No one had an inkling that there could be anything smaller. Then the microscope was invented.

No one knows quite who invented it. Craftsmen had probably been using drops of water and rock crystals to magnify their work for thousands of years, but the credit for creating a special apparatus to magnify usually goes to a Dutch spectacle maker of the late sixteenth century called Zacharias Janssen. He was skilled in grinding glass to make it magnify, and his breakthrough was to put two glass lenses together to dramatically increase the magnification.

Scientists soon began to catch on to the possibilities of Janssen's microscope and to make improvements to it. In 1665, the great English scientist Robert Hooke published a book called *Micrographia* in which he explained the basic principles of microscopy – and which contained the first drawing of a tiny living cell, which he had seen in a slice of cork. But it never occurred to the scientists of the day to look for life with a microscope anywhere where it could not be seen with the naked eye. They simply used their microscopes to study such things as skin in close up or cork or hairs.

Leeuwenhoek's brilliant insight was to use his microscope to look in all kinds of places where he could not see life with the naked eye, particularly liquids, and he studied an extraordinary range of things with his microscope – raindrops, tooth plaque, dung, sperm, blood and much more besides. It was here in these apparently lifeless substances that Leeuwenhoek discovered the teeming richness of microscopic life.

The humble draper

Leeuwenhoek was born in Delft on 24 October 1632 of quite humble origins. His father was a basket maker and his mother's family were brewers. Very little is known about his childhood. He walked to school in the nearby town of Warmond, and then, when he was 16, he became an apprentice at his uncle's linen drapery. Four years later he returned to Delft, set himself up as a draper and remained so for the

A light micrograph of a colony of *Conochilus hippocrepis* rotifer worms. The colony is composed of between 50 and 100 individual worms, attached at the base of their feet, with their bodies radiating outwards from a common centre, as seen here.

rest of his 91 years. He was also appointed chamberlain to the Delft Council in 1660, and was one of the trustees of the estate of his friend the artist Jan Vermeer.

He married his first wife, Barbara, when he was 22. They had five children before Barbara died 12 years later, but only one of them – his daughter Maria – survived. He married again, but his second wife died, too, and when she did, Maria moved in with her father and cared for him with utter devotion for the rest of his life. The little household in Delft was completed by a long-haired dog, a talking parrot and a quiet horse.

Home-made microscopes

Leeuwenhoek was probably inspired to take up microscopy sometime around 1668 after seeing a copy of Hooke's *Micrographia*, which was very popular at the time, though he may already have been using lenses to examine cloth. Unlike Hooke, Leeuwenhoek did not use a two-lens 'compound' microscope, but a single, high-quality lens, which could be described simply as a magnifying glass rather than a microscope. At the time, such simple microscopics were much easier to get a clear picture with. Problems with blurring meant it was impossible to make compound microscopes that magnified more than 20 or 30 times. But Leeuwenhoek was able to grind his own single-lens microscopes, and after years of honing his technique – and with the help of his acute eyesight and careful lighting – he was able to see things clearly at magnifications of well over 200 times, something no one else managed to achieve for almost two centuries.

Leeuwenhoek is known to have made over 500 of these single-lens microscopes, but only 10 survive. They are very simple devices, just a few inches long, with the lens mounted in a tiny hole in a brass plate. The specimen is mounted on a point that sticks up in front of the lens. Two screws move the specimen for focusing. All else that is needed is careful lighting and a very steady, sharp eye.

Letters to the Royal Society

After a few years of perfecting his little microscopes, Leeuwenhoek began examining such things as mould and insects in incredible close-up, seeing for the first time the complex structure of a bee's eye, for instance. In 1673, Leeuwenhoek contacted the Dutch physician and anatomist Regnier de Graaf to tell him what he had found. Although just 32, de Graaf was already famous for his discovery of the egg-making sites or follicles in the human ovary which now bear his name. De Graaf wrote to Henry Oldenburg, the president of the Royal Society in London – the hub of seventeenth-century science – about his fellow Dutchman and his marvellous microscope. Because of de Graaf's reputation, Oldenburg at once invited Leeuwenhoek to write a letter to the Society reporting his findings, to be published in *Philosophical Transactions*, the Society's journal.

For the humble Dutch draper, this invitation was undoubtedly rather overwhelming. In the letter accompanying the report, he wrote that he had never tried to publish his results because he was not sure that he could express himself effectively. Nor could he speak Latin, the international language of scientists at the time. He wrote in Dutch and got a local man to translate it into Latin. Later letters were sent in Dutch and translated into English.

Leeuwenhoek's early observations were nothing startling, and this first letter reported no more than what Hooke had seen a decade earlier. But it was well enough received, and Leeuwenhoek was encouraged to continue with his work and go on sending letters to the society. In fact, he went on sending them for 50 years, and he sent more reports than any member of the society, before or since.

Animalcules

Although Leeuwenhoek's first letter had not reported anything dramatic, he was perfecting his technique. Then, in 1674, he reported seeing little creatures in lake water:

> *I found floating therein divers earthly particles, and some green streaks, spirally wound serpentwise, and orderly arranged after the manner of copper or tin worms which distillers use to cool their liquors as they distil over. The whole circumference of each of these streaks was about the thickness of a hair of one's head.*

The most remarkable letter was Letter 18, the letter of October 1676, in which he reports the discovery of what we now call bacteria, in drops of water (see box on p. 48). But his discoveries did not stop there. In 1677, he described how he examined his own semen. This too was swarming with the little animals we now call sperm. But unlike the creatures in rainwater, the animalcules (*lit.* tiny animals) in semen were all identical. Each of the many thousands he looked at had the same tiny tail and head, and nothing else. He could see them swimming like tadpoles in the semen. This was too much for others to believe, and it was decades before Leeuwenhoek's description of sperm was accepted.

A mouthful of life

In Letter 39 of 1683, Leeuwenhoek reports how he examined his own saliva and plaque scraped from his teeth. He says that the saliva contains no animalcules, but the plaque is teeming with them, 'very prettily amoving'. 'The biggest sort,' he says, 'had a very strong and swift motion, and shot [about] like a pike through the water. The second sort ... oft-times spun around like a top ... and those were far more in number.' He also says that plaque taken from his teeth after he began drinking scalding coffee contained no animalcules, and concludes: 'The heat of that coffee probably killed my little animals.' He confirmed this when he found animalcules in plaque taken from his back teeth, not exposed directly to the coffee.

A little later Leeuwenhoek came close to realizing that bacteria can be germs that cause disease when he noted a dramatic increase in the numbers of animalcules in his mouth when he was ill, and in a rotting tooth. People have sometimes been disappointed that Leeuwenhoek did not go one step further and identify bacteria as germs. It took another century before Louis Pasteur made that step. But in many ways, Leeuwenhoek's understanding was closer to our modern view of bacteria and the important, often beneficial role they play in the world, as well as germs.

By the time he died on 26 August 1723, Leeuwenhoek was modestly famous. The Holy Roman Emperor had visited his house to see his microscopic marvels, and so too had Queen Mary of England and various other notable people. But he never sought anything more than the quiet life with his daughter and microscope. Several years before his death, he made a beautiful wooden cabinet to hold his best microscopes and specimens. When he died, his daughter Maria sent this to the Royal Society as he had requested, and there it stayed for a century before mysteriously disappearing – only for various items from it to reappear in recent years. Despite his achievements, Leeuwenhoek was quite quickly forgotten, and it was his devoted daughter alone who later erected a tiny monument to him in Delft, commemorating his great discoveries.

Discovering bacteria

Leeuwenhoek's famous Letter 18 was seventeen and a half pages long and begins modestly enough as he goes through his scientific diary for the previous year: 'In the year 1675, about half-way through September ... I discovered little creatures in rain which had stood but a few days in a new tub that was painted blue within.' Leeuwenhoek thought that it was worth exploring what else there might be in water for, as he said, 'these little animals to my eye were more than ten thousand times smaller than ... the water flea or water louse, which you can see alive and moving with the naked eye'.

He describes how he went on examining water from different sources – rainwater, pondwater, well water, sea water, and so on, each left exposed to the air before he examined them. Each time he saw some of these incredibly tiny creatures. He was especially interested in those that seemed to have 'legs' and 'tails' that allow them to scurry about in the water: 'When [some of] these animals bestirred themselves, they sometimes stuck out two little horns, which were continually moved, after the fashion of a horse's ears.'

It is not entirely certain what these creatures were, though we might guess they are bacteria. But then Leeuwenhoek describes how he examined water that had been infused with ground pepper, and what he saw there were clearly bacteria. These little creatures moved so little he was not always sure they were alive, but there were enough clues.

The idea that there were tens of thousands of tiny creatures in a single drop of water was so astounding that Henry Oldenburg asked Leeuwenhoek to provide independent witnesses to verify his findings. The draper invited several of Delft's most respectable citizens, including the vicar, to look into his microscope. They confirmed what he saw. A year later, Hooke too confirmed the findings with his own observations, performed in front of expert witnesses that included Christopher Wren, the architect of London's St Paul's Cathedral.

Robert Hooke

Robert Hooke (1635–1703) is one of the great unsung heroes of science. He made crucial contributions to almost every branch of the scientific revolution of the seventeenth century, yet remains a shadowy, little-known figure.

1635–1703

MOST PEOPLE KNOW OF HOOKE'S NAME ONLY BECAUSE OF THE LAW OF ELASTICITY – how things stretch or squeeze – which bears his name. This basically says that the force with which any springy material bounces back when squeezed or stretched depends on the force used to squeeze or stretch it. This is an important law, but it is just one tiny example of Hooke's overall contribution to science.

The sheer range of Hooke's activity is quite astonishing, and some experts have described him as the English Leonardo da Vinci. He was perhaps the most practical and inventive of all the great scientists, and his range of inventions matches, if not surpasses, his theoretical ideas.

Among Hooke's many inventions were the ear trumpet, the first practical spirit level, sash windows, balancing escapements in clocks, the anenometer (for measuring wind speed), the hygrometer (for measuring humidity), the wheel barometer (for showing how air pressure varies), the crosshair (for sights in telescopes and, later, gunsights), the iris diaphragm (later the aperture in cameras), micro-dots for secret messages, respirators for easing breathing, a diving bell, an air pump, the universal joint (now widely used in car drive shafts), a self-stabilizing keel for boats, an early form of sonar, mercury amalgam (later used for dental fillings), the micrometer, an airgun – the list goes on and on. And like Leonardo, Hooke worked on flying machines and even anticipated the coming of the steam engine. Not for nothing did his friend, the famous diarist John Aubrey, describe him as 'certainly the greatest mechanick this day in the world'.

But Hooke was by no means just an ingenious technician. He was also a visionary scientist. He is known to have collaborated with most of the great scientists of his day – Boyle, Newton, Huygens, Leeuwenhoek, Wren and many others – and clearly contributed a great deal to their work. Much of the record of Hooke's work is lost, so his contribution was not always clear or recognized, even at the time. Some, like Boyle, freely acknowledged Hooke's valuable role in helping formulate his famous gas laws. Newton, however, fought long and bitterly with Hooke over just who had come up with the crucial idea that the force of gravity diminishes as a square of the distance between things (the inverse square law). The evidence points in Hooke's favour, but Newton effectively obliterated the record of Hooke's role in his work.

In some areas, though, Hooke's contribution is incontestable. He was the great pioneer of microscope studies, coining the word 'cell' for the microscopic packages from which all living things are built. His book *Micrographia* was the greatest book of the age on the microscopic world – and an inspiration for men as diverse as Leeuwenhoek and the diarist Samuel Pepys. He was also the founder of the science of meteorology, showing how to measure atmospheric quantities like wind speed and air pressure, and suggesting how accurate record keeping could lead to weather forecasting.

As if all this wasn't enough, Hooke was a major architect, not only collaborating with Sir Christopher Wren on the rebuilding of London after the Great Fire of 1666, but designing many buildings in his own right. The famous Monument to the fire in London, the world's tallest Greek-style column, is thought to have been designed by him. So too was the great palace-like Bethlehem Hospital (now destroyed), and maybe the famous Whispering Gallery of St Paul's Cathedral too.

Young Hooke

Robert Hooke was born on 18 July 1635 at Freshwater on the Isle of Wight. He later described his early childhood as charmed and trouble free, but it was in fact marred by smallpox, which left him scarred for life. His mother is nowhere mentioned in later recollections, and he seems to have spent a lot of time alone, wandering the chalk landscape of the Isle of Wight, which is how he may have gained his first knowledge of fossils.

The great tragedy of young Robert's childhood happened when he was just 11 years old. This was the suicide of his clergyman father, who lost his living in the wake of the royalist defeat in the English civil war. Robert was left with just a small legacy of £40. Maybe with the help of neighbours, or maybe entirely on his own initiative, he travelled to London to enrol as an apprentice in the studio of England's most accomplished painter, Peter Lely, who was famous for his portraits of Charles I. After a year or so, in which he learned valuable drawing skills, he decided he needed a proper education and used his legacy to enrol in Westminster school. The school's formidable headmaster, Richard Busby, recognized he had someone special on his hands when the 13-year-old Robert read all of Euclid's works in his first week.

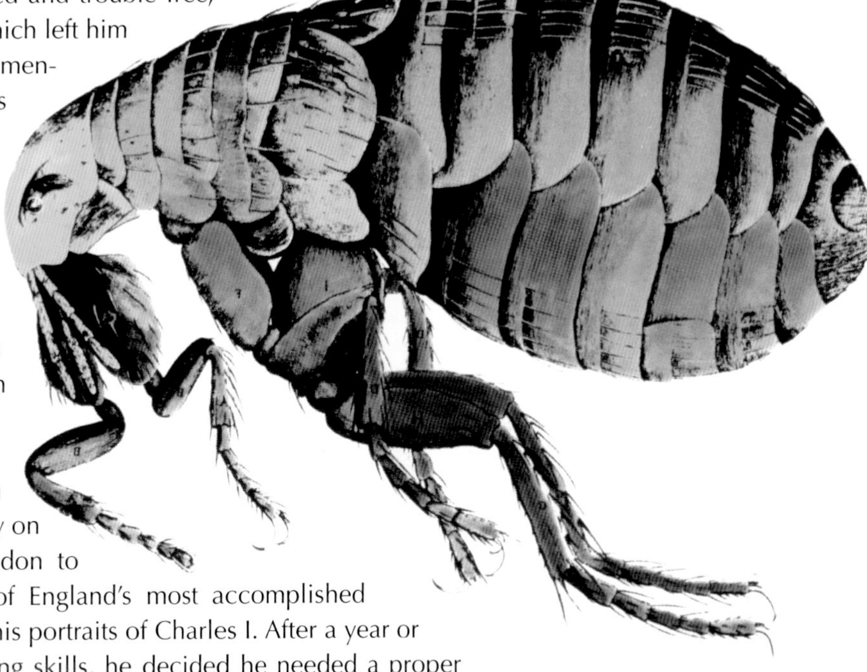

A drawing of a flea by Robert Hooke and published in Micrographia *in 1665.*

Busby became the young Hooke's patron and helped ensure he progressed to Oxford University, where he paid his way with a choral scholarship. It was while at Oxford in the late 1650s that Hooke met the famous Irish chemist Robert Boyle, and for a while he became Boyle's assistant. At Boyle's request Hooke came up with the ingenious design for an air pump which allowed Boyle to show how, as a gas is compressed, its pressure increases in proportion, and so formulate his famous gas laws. Hooke's design provided the basis for the design of air pumps even today.

The Royal Society

Boyle and Hooke formed the nucleus of brilliant scientists at Gresham College in Oxford who were to create the world's most famous scientific society, the Royal Society. When the Society was launched in 1662, Boyle ensured that Hooke was appointed its first curator of experiments. This post, which for many years was unpaid, meant that Hooke was expected to give practical experiments at every meeting of the Society. It is hard to imagine anyone but Hooke being willing to fulfil such an arduous brief – or making such an astonishing success of it. In response to the Society's demands, Hooke came up with a greater wealth of original scientific ideas and practical demonstrations than anyone before or since. It seemed to suit Hooke's agile mind to leap from one task to another with amazing rapidity, but it may be why he never developed any one single idea to the point where he became lastingly famous for it.

All the same, Hooke achieved worldwide fame when he published the fruits of his experimental work with microscopes in the book *Micrographia* in 1665 (see box on p. 57). But while he continued to contribute a seemingly endless stream of ideas to almost every field of science, he began to become increasingly bitter and isolated as other scientists started to take the credit for theories he had either suggested or proved with practical demonstrations.

The battle with Newton

The bitterest dispute was that with Newton. When Newton made his theory of light and colour known in 1672 (see p. 61), Hooke pointed out that what was right in Newton's theory had been suggested by him seven years previously. He also argued that light did not travel as particles, but as waves – an idea later credited to the Dutch scientist Huygens (see p. 49). The battle lines began to be drawn between these giants of science.

When Newton first mentioned his ideas on gravity in the 1680s, he had suggested that gravity was a constant – that is, the same strength everywhere. Hooke suggested that the elliptical course of the planets' orbits might be better explained if gravity was not constant, but obeyed an inverse square law, diminishing as the square of the distance between objects. But of course Hooke had as yet no proof of his idea – even though a decade earlier he made important observations to prove it using a zenith telescope from his rooms in Oxford. In response to Hooke's publication, Newton wrote his great book *Principia* over the next few years, in which he provided the mathematical proofs that gravity does indeed obey this inverse square law.

As Newton came to the point of publishing *Principia*, Hooke insisted to the Royal Society that he be properly credited with this key theoretical cornerstone of Newton's work. Newton was incensed and immediately deleted virtually all references to Hooke in his book. In a famous reposte, Newton quoted an old Latin rhetorical tag: 'Merely because one says something could be so, it does not follow that it is so' – implying that the credit should go entirely to the scientist who proved the idea. Newton was by this time such a powerful figure that his vindictive view of Hooke stuck – and even today most people still go along with Newton's uncharitable view that the dilettante Hooke lacked the solid science and mathematics to convert the glimmerings of a theory into a solid, proven fact.

Micrographia

When Hooke published his great book *Micrographia*, it caused an immediate sensation and was perhaps the first popular science book in history. The famous diarist Pepys obtained a copy as soon as he could, and wrote, 'Before I went to bed I sat up until two o'clock in my chamber reading Mr Hooke's Microscopical Observations, the most ingenious book that I ever read in my life.'

Hooke was the probably first scientist to use the microscope to study life in microscopic detail. Using a first-class compound microscope that he made for himself, he studied organisms as diverse as insects, sponges and foraminifera. *Micrographia* is a meticulous and detailed record of his observations, and includes his own superb drawings of what he saw through his microscope. He was the first to see clearly such things as the scales on a flea's body and the minute hairs on a fly's legs.

His most famous observation was of thin slices of cork, of which he wrote: 'I could exceedingly plainly perceive it to be all perforated and porous, much like a honeycomb, but that the pores of it were regular. These pores, or cells ... were indeed the first microscopical pores I ever saw, and perhaps, that were ever seen ...' In fact, as we know now, Hooke had discovered living cells, which he named because they reminded him of the cells in a monastery.

Hooke also studied fossils and noted the similarities between petrified wood and fossil shells and living wood and shells, and went on to make the first description of the process of fossilization, in which minerals gradually replace living tissues to turn dead organisms to stone.

He later went even further, suggesting that seashell fossils found in high mountains indicate that in the past the world was subjected to massive earthquakes which threw up ancient seabeds to form mountains. Hooke even thought that some of these fossils might be of species of creature that no longer existed. Such ideas were way ahead of their time, and only came to prominence in the great revolution which transformed geology and formed the basis of Darwin's theory of evolution almost 200 years later.

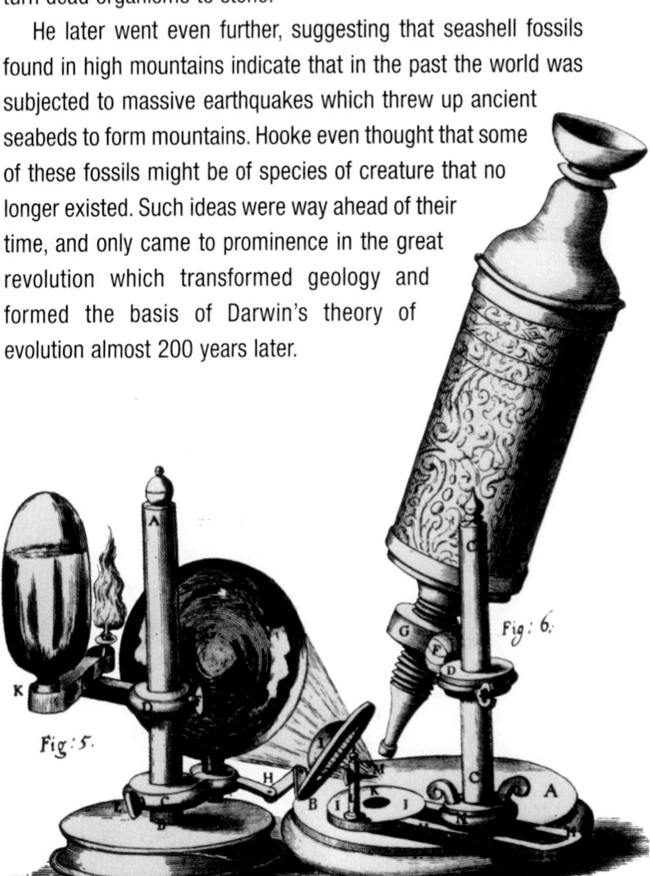

A microscope designed by Robert Hooke in 1665.

Final years

As Hooke grew older, he became increasingly depressed and withdrawn. This was not enough for Newton, though. So deep was Newton's hatred of Hooke, it is said, that when Newton became president of the Royal Society, he destroyed the portraits of Hooke that hung there.

By the time he reached his 60s, Hooke was a physical and emotional wreck. Some of this was undoubtedly down to the stress of his battle with Newton, but a lot was self-inflicted. He had lived at a punishing pace all his life, working through night after night on experiments. He had tried many experiments on his own body, and they had taken their toll. Moreover, to keep up the impossible pace he set himself, he had stuffed himself again and again with stimulant drugs. It couldn't last. His final months were a time of physical deterioration, and of isolation as one by one his friends fell away, unable to cope with his bouts of paranoia. He died alone in his Gresham College rooms on 3 March 1703, tended only by a young maid.

Sir Isaac Newton

By universal consent, Isaac Newton (1642–1727) was the greatest scientist of all time. He is most famous for 'discovering' gravity and for his three great laws of motion, which underpin all modern physics. But he also created the mathematics of calculus, unlocked the secrets of coloured light and much more besides.

1642–1727

WE TAKE NEWTON'S VIEWS OF THE WAYS THINGS MOVE so much for granted nowadays that it is hard to imagine just how revolutionary they were Newton's day, and just what an extraordinary breakthrough they were. Before Newton, there had been no notion that the movement of the fish in the sea or papers disturbed by a breeze had anything in common whatsoever with the movement of the heavens, let alone that they were predictable in any way. They were seen to be controlled by unique, local factors, or else by the whim of the gods. The universe was, essentially, a mysterious, capricious place.

With his law of gravity and his three laws of motion, Newton showed that every movement, large or small, on the ground or in the furthest reaches of space, behaves according to the same simple, universal laws. In his book *Philosophae naturalis principia mathematica* (The mathematical principles of natural philosophy), or simply *Principia*, the greatest science book ever written, he suddenly blew away the universe's chaotic mystery and showed that everything everywhere behaves in an orderly, entirely understandable way. It was as if the whole universe had been revealed at last as some great incredibly complex clockwork machine, and Newton's laws were the key to its working. Incredibly, it was shown that the laws worked out with experiments in laboratories here on the ground can be applied right across the universe.

Even more significantly, the *Principia* showed how every single movement in the universe can be analyzed mathematically, and Newton provided the mathematical tools to do it, with the two branches of mathematics that he created – differential and integral calculus (see p. 61). Armed with Newton's laws and Newton's mathematics, it became possible not just to work out what is going on in any movement, from the lifting of a coffee cup to the orbit of a planet, but to predict what would happen when, for instance, a train goes across a new bridge or when a spacecraft is launched. Newton made it possible, in theory, to predict the movement of everything in the universe forever, from the greatest star to the tiniest molecule.

No wonder then that as Newton's ideas began to be widely understood, he came to be regarded with awe in the eighteenth century. His revelation that the universe behaves according to predictable, universal laws ushered in a new, optimistic age, the age of Enlightment, in which people believed that humankind can learn to understand and improve the world. As the poet Alexander Pope said in his famous satirical verse, 'Nature and nature's laws lay hid in night. God said, "Let Newton be!" and all was light.'

If this optimism seems clouded today, it is only by doubts that we will do the right thing – not that things are ultimately beyond our understanding in the way Newton initiated. And if Einstein's discoveries have shown a subtler, deeper insight as to how universal laws work at extremes (see p. 119), Newton's laws underpin our basic understanding of how things work on an everyday scale.

The young genius

Isaac Newton was born on Christmas Day 1642 in a small manor house in the Lincolnshire village of Woolsthorpe. He was premature, 'so little they could fit him into a quart pot', and so sickly he was not expected to survive. In fact, he proved remarkably healthy and lived 84 years. His father was already dead by the time Newton was born. When he was just 18 months old, his poor widowed mother married a wealthy old local minister in nearby North Witham, but left the infant Isaac with his grandparents.

It may be that Isaac never recovered from this early abandonment. Even though his mother returned home to her son when her new husband died seven years later, Isaac later confessed that he remembered 'threatening my (step)father and mother to burn them and their house over them'. Throughout his life, Newton carried a terrible suppressed anger and sense of resentment that made him a very difficult man to deal with.

The introverted Isaac went to school at the age of 12 but showed no signs of any intellectual prowess until he was bullied one day at school. In a towering rage the young Newton fought back until his larger opponent was a quivering wreck. But

Isaac Newton conducts his famous experiment on light, using a prism to refract a ray of light from a hole in the shutters over a window. A white surface (far left) allowed Newton to observe that light splits into the different colours observed in rainbows.

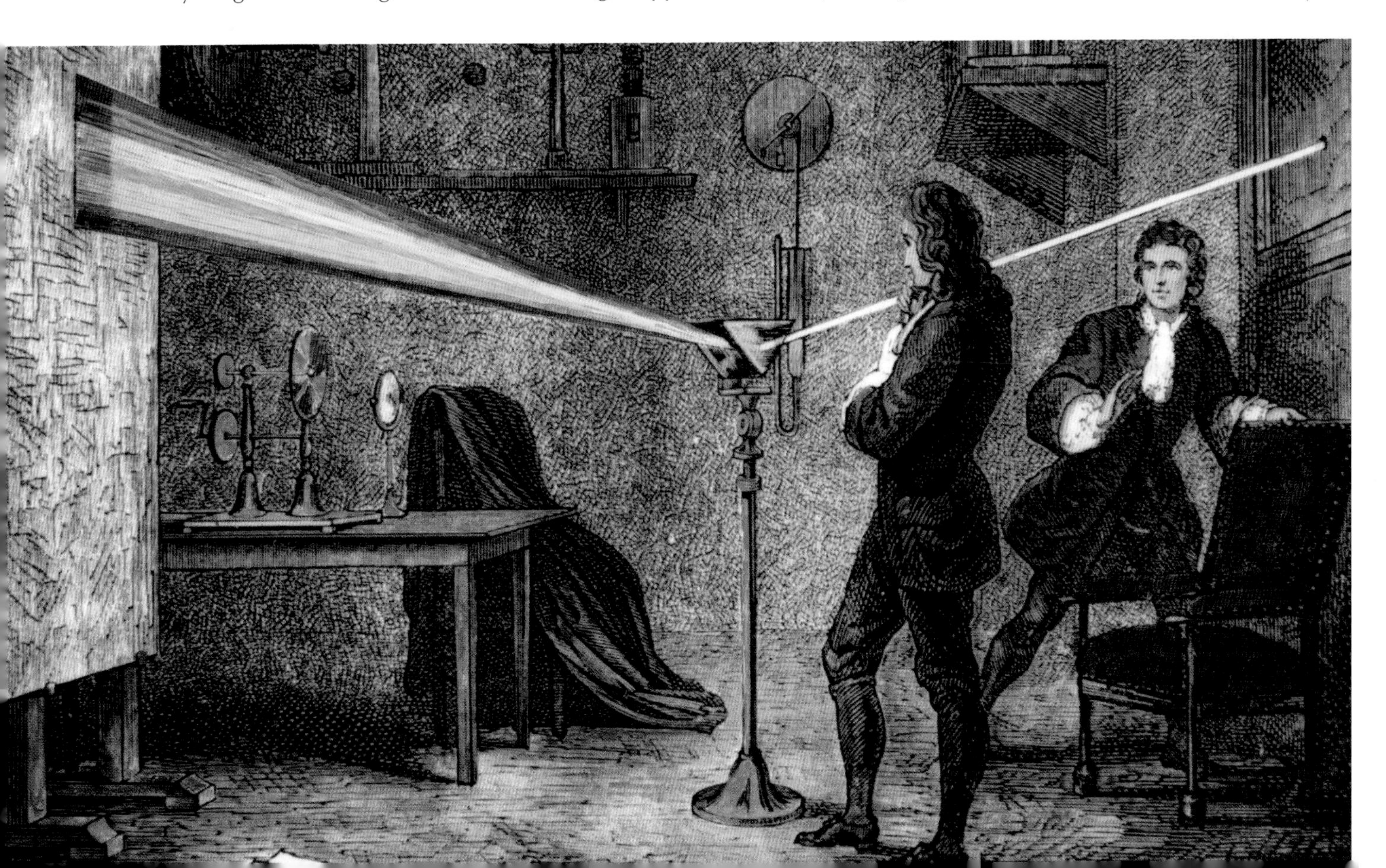

Newton's great laws

The story goes that the inspiration for the idea of gravity came to Newton one late summer day in 1666 as he sat thinking in the garden at Woolsthorpe, and saw an apple drop from a tree. This is often dismissed as legend, but Newton himself claimed it was so.

It is not entirely clear what the falling apple made him think, but Newton's real insight was to understand just why it fell. In the previous half century, Kepler had shown that planets have elliptical (oval) orbits, and Galileo had shown that things accelerate at an even pace as they fall towards the ground. Yet no one had thought of connecting these two events, let alone showing they have the same universal cause.

Newton realized that the apple was not just falling but being pulled by an invisible force – and later wondered if this same force might be holding the planets in orbit. Just as gravity pulls the apple to Earth, so gravity keeps the Moon in its orbit round the Earth and the planets round the Sun, and stops them flying off into space. From this simple but brilliant idea, Newton developed his theory of gravity, the universal force that tries to pull all matter together. With mathematical proofs, he showed that this force must be the same everywhere, and that the pull between two things depends on their mass (the amount of matter in them) and the square of the distance between them.

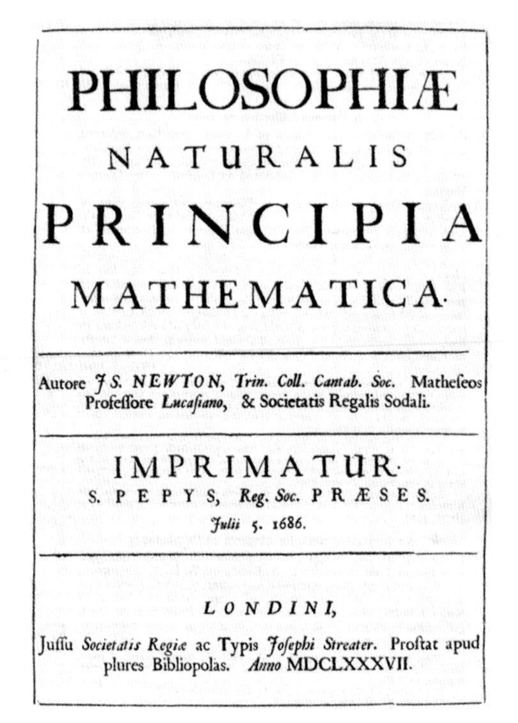

PHILOSOPHIÆ
NATURALIS
PRINCIPIA
MATHEMATICA.

Autore *J. S. NEWTON*, *Trin. Coll. Cantab. Soc.* Matheſeos Profeſſore *Lucaſiano*, & Societatis Regalis Sodali.

IMPRIMATUR.
S. PEPYS, *Reg. Soc.* PRÆSES.
Julii 5. 1686.

LONDINI,

Juſſu *Societatis Regiæ* ac Typis *Joſephi Streater*. Proſtat apud plures Bibliopolas. *Anno* MDCLXXXVII.

The title page of Newton's masterwork, generally known as Principia Mathematica *(1687).*

Over the next 20 years, Newton refined his idea of gravitation into a comprehensive system that included the 3 great laws of motion. The first law was the idea of inertia or momentum. It basically means that things keep moving at the same speed in a straight line unless something pushes or pulls on them – that is, a force. He applied this to the Moon, showing that the Moon tries to carry on in a straight line, but gravity pulls it into an orbit. The second law is the idea that the rate and direction of any change depends entirely on the strength of the force that causes it, and how heavy the affected object is. If the Moon were closer to the Earth, the pull of gravity between them would be so strong that the Moon would be dragged down to crash into the Earth. If it were further away, gravity would be so weak that the Moon would fly off into space. The third law showed that every action and reaction are equal and opposite, so that when two things crash together they bounce off one another with equal force.

Newton did not stop there. He was determined to humiliate his opponent in the classroom too. Soon Newton became deeply involved in his academic pursuits, especially science, and amazed the locals with such things as handmade water clocks and flying lanterns.

Recognizing the boy's talents, Newton's teacher, John Stokes, and his uncle, William Ayscough, encouraged him to try for Cambridge, and he was accepted for a place there in 1661 when he was 19. By this time, Newton was so concerned to follow up his own research that he barely bothered with the course work and was almost failed. Unknown to his tutors, though, he was already going far beyond them, developing the latest revolutionary mathematical and scientific ideas of the French genius René Descartes which were only just beginning to filter into England.

The miraculous year

In August 1665, plague struck right across southern England. Cambridge was effectively closed and Newton was forced to return to Woolsthorpe, where he remained for a year. This enforced retirement turned out to be a truly astonishing year. In the peace and quiet of his Lincolnshire home, Newton developed his work on Descartes to create the mathematics of calculus, which studies how fast things change – an essential for understanding acceleration, which is the way forces work. Calculus, which Newton called the method of fluxions, took Archimedes's brilliant way of using polygons and rectangles to work out the areas of circles and curves, and pushed it one giant step further, to show how the tangent or slope of any point on a curve on a graph can be analyzed, such as one showing time against distance travelled. That way, the speed or acceleration at any given moment can be analyzed.

Newton's greatest discoveries were the law of gravity and the laws of motion (see box on p. 60). Then he added a third. While at Stourbridge market, he picked up a pair of glass prisms, and he began to research the nature of light and colours. By using the prism to split daylight into the colours of the rainbow, and then using another to recombine them into white light, he showed that white light is made up of all the rainbow's colours (see box on p. 62).

Remarkably, of all his scientific discoveries during his time in Woolsthorpe, this was the only one he revealed when he returned to Cambridge the following year. It was enough, though, to earn him the post of Lucasian professor of Mathematics in 1669. Newton remained secretive all his life, writing reams of research findings, but only showing them when challenged or when someone seemed about to come to the same conclusions as him.

The turbulent scientist

This secrecy was to cause him tremendous problems in the future. It took him 30 years to publish his work on calculus, for instance. In the intervening period, the German philosopher Liebnitz had published his own independently discovered version and given it the name calculus, which stuck. The row over who had got there first became so bitter that the Royal Society held an inquiry to sort out the mess. Newton hijacked the committee's report to ensure it found in his favour, which it would have done anyway. Even so, Newton's furious tirades against Liebnitz went on even after the German philosopher's death.

'Nature and Nature's laws lay hid in night God said "Let Newton be!" and all was light'

Similarly, he was only stung into publishing his ideas on gravity and motion when his only serious scientific rival in England, Robert Hooke, claimed in 1684 that he had solved the planetary motion problem with an inverse square law that governed the way the planets moved. Hooke was right about the inverse square law, but he had no notion of why it worked, or even how to prove it. It was in his determination to put Hooke in his place that Newton realized that his idea of gravity and laws of motion – which he had so far only applied to Earth – might work for the heavens too. He sat down to work on his masterpiece, the *Principia*, which he finished 30 months later.

By this time, Newton was already famous for a remarkable telescope he had made in 1668. Telescopes with lenses were already well developed and getting quite big in Newton's day. The problem was that the bigger they got, the more they suffered from distorted colour fringes as the light rays were bent through the glass of the lens. Newton solved the problem by swapping the lenses for curved mirrors so that the light rays did not have to pass through glass but simply reflected off it. This also doubled the light back on itself, making the telescope much more compact. A 'Newtonian' telescope could be just 15 cm long yet as powerful as a lens telescope a metre long – and avoided all the problems of colour fringing.

Newton on light

Newton's prism reflects spectral light onto the manuscript of his work Opticks.

In his 1704 book *Opticks*, Newton suggested that a beam of light is a stream of tiny particles or 'corpuscles', travelling at huge speed. If so, he argued, this would explain why light can travel though a vacuum, where there is nothing to carry it. It also explained, he believed, why light travels in straight lines and casts sharp shadows – and is reflected from mirrors, the tiny particles bouncing off just like tennis balls off a wall. He thought the bending or 'refraction' of light might be caused by the corpuscles travelling faster in glass or water than air. At around the same time, the Dutch scientist Christian Huygens came up with an equally convincing but wholly contradictory theory that light travels in waves like ripples on a pond, and not as particles. Between them they started a debate which is still not completely resolved today. Current thinking suggests both are right at different times or in different situations.

Newton's fame

It was the invention of this telescope that got Newton elected to the prestigious Royal Society, the leading scientific society in Europe and a gathering place for all the finest minds in Britain. But almost as soon as he had been elected, he resigned after a bitter row with Robert Hooke over papers that Newton wrote detailing his work with prisms, which Hooke wrongly criticized. Newton vowed never to return to the Society while Hooke was there. It was only when Hooke died in 1703 that Newton returned as president, which he remained until he died in 1723.

The *Principia*, published in 1687, caused a sensation and made Newton internationally famous. Indeed, he was the first scientific superstar. He was elected as MP for Cambridge and, when he came to London to do his sessions in Parliament, he found himself feted by everyone from the king to the great philosopher John Locke, and surrounded by young acolytes. One of these acolytes was a gifted young Swiss mathematician called Fatio de Duillier. The 48-year-old Newton is said to have fallen in love with Fatio. When Fatio left for Switzerland three years later, Newton was distraught.

There is no way of knowing if this is merely coincidence, but from this time on Newton never embarked on any more scientific work. For a while he became increasingly reclusive, concentrating on the alchemical research (see box) that he regarded as, if anything, more important than his scientific research.

Newton the alchemist

Isaac Newton is generally considered to be the first great modern scientist, and his book *Principia* as the first of all the most important scientific achievements of the last 350 years. It was a product of a science that prides itself on observation, experiment and pragmatic, honest logic. Yet in some ways he was also the last great magician of the medieval age. He spent more than half of his life avidly researching alchemy and astrology, spending day after day and night after night secretly working in his laboratory trying to turn base metal into gold, or scanning ancient texts for insights. This work was more important to him even than his science. He was not content with his discoveries of the workings of the physical world, but wanted to discover as well the mechanics of human life. Most of his notes on the subject were destroyed in a fire started by his dog Diamond, so only a fraction of his huge volume of research survives – and much of that is now completely unintelligible to us. Although we are apt to dismiss all this work as peculiar mumbo-jumbo not worthy of such a fine mind, it was his acceptance of the possibility that there are mysterious forces in the world that led him to the idea of an invisible gravitational force – something that the more rationally minded Galileo had not been able to accept.

The final years

In 1696, Newton was invited to join the Royal Mint. The official story was that the position was a reward for his intellectual achievements. The famous French philosopher Voltaire believed it was simply because he had a pretty niece married to the right person. Whatever the truth, Newton embarked on his job with fanatical zeal, determined to stamp out the counterfeiting of coins that was then undermining England's currency.

Newton remained in the post of Master of the Mint for the rest of his life. After Hooke died, Newton combined it with the post of president of the Royal Society, a task that he pursued with the same vigour, though he was now in his sixties.

Newton died on 20 March 1723. When he was buried with a grand funeral at Westminster Abbey, attended by all the great dignitaries of the day, huge crowds came to watch the cortege pass by. Voltaire was visiting London at the time and wrote, 'England honours a mathematician as other nations honour a king who has done well by his subjects.' It was true.

Carolus Linnaeus

Sometimes known in Sweden as the Flower King, Carolus Linnaeus (1707–78) was the great eighteenth-century botanist who devised the system all scientists use today for classifying living things into species and classes with a two-part Latin name.

1707–78

In the seventeenth century, botanists and zoologists from Europe were just beginning to discover the incredible diversity of natural life in the world. Some were beginning to look at the plants and animals around them with the close interest inspired by the scientific revolution of the age. Others studied exotic plants and animals brought back from the distant places ships were now visiting.

The more botanists and zoologists looked at nature, or rather Creation, as it was often called, the more confusing its teeming range seemed to them. When people wrote books or treatises on animals, there seemed to be no other way of classifying them but alphabetically, and there was no way of distinguishing the real from the mythical. Books on animals, or 'bestiaries', might begin with 'Antalopes', move on to 'Apes', and then 'Areopathogus'. This might be the order in English – but the order in each language would of course be different.

The first great attempt to sort out this chaos was made by the English botanist John Ray (1627–1705). In 1671, Ray went on an extensive specimen-gathering trip through Europe with his zoologist friend Francis Willoughby. Unfortunately, Willoughby died shortly after their return, but Ray carried on their work. Ray came up with a scheme for classifying all plants and animals. His brief *Methodus Plantarum* (1682) provided the first definition of a species as 'a set of individuals who give rise through reproduction to new individuals similar to themselves'. In his *Historia Plantarum*, Ray, for the first time, grouped species scientifically according to their structure.

Linnaeus took up where Ray left off to create his definitive 'system of nature' half a century later. Linnaeus was not alone in trying to find a classification system. By 1799, over 50 schemes had been proposed, but Linnaeus's system had two key features which guaranteed its survival.

First of all, Linnaeus grouped plants according to their sexual organs – that is, the parts of a plant involved in reproduction. Secondly, he gave each species a two-part Latin name, such as *Linnaea borealis*, the marsh twinflower, part of the honey suckle family, and named after the great botanist himself. The first part always refers to the name of the group it belongs to, and the second part is the species name. This system was so powerful and effective that it was adopted by botanists around the world by the end of the century and has remained in place ever since.

A Swedish childhood

Carolus Linnaeus was born in 1707 by the shores of Lake Möckeln in southeastern Sweden. Later in life, he recalled it as one of the most beautiful places in Sweden: 'When one sits there in the summer and listens to the cuckoo and the song of all the other birds, the chirping and humming of the insects; when one looks at the shining, gaily coloured flowers; one is completely stunned by the incredible resourcefulness of the Creator.'

Yet at school, young Carolus showed so little interest in either botany or theology that his father thought to apprentice him to a shoemaker. Fortunately, a perceptive schoolmaster suggested he go to Uppsala University as a medical student.

At Uppsala, Linnaeus was immediately captivated by the demonstrations performed by the ageing botanist Olof Celsius in the university's botanical gardens. Indeed, Linnaeus's interest became so avid that in 1732, when he was 25, he was sent by the Uppsala Science Society to gather specimens in Lapland. He was thoroughly delighted with what he found there, and his finds included the small white Arctic flower *Linnaeus borealis* (Linnaeus of the north), which became his trademark. The most famous portrait of him shows him wearing the white flower and festooned with traditional items of the Sami people of Lapland. His findings there were later published in his book *Flora Lapponica* (1737).

Cataloguing creation

When he returned south, Linnaeus went to the Netherlands to finish his medical studies. Whilst there he studied the incredible range of plants in the garden and herbarium of the wealthy banker George Clifford. He also met Peter Artedi, another young and enthusiastic naturalist. Together they conceived a plan to classify all creation. Artedi was to study the fishes and land animals and Linnaeus, birds and plants. Sadly, Artedi fell into an Amsterdam canal and drowned, so it was left to Linnaeus to continue the project alone.

Linnaeus's basic scheme was sketched out in a little pamphlet called *Systema Naturae* (*Systems of Nature*), which he published in Holland in 1735. 'In these few pages,' Linnaeus explained, 'is handled the great analogy which is found between plants and animals, in their increase in like measure according to their kind, and what I have here simply written, I pray may be favourably received.'

Plant sex

What Linnaeus was talking about with his talk of the analogy between plants and animals was the sexual nature of plants. A few decades before, German botanist Rudolph Camerarius (1665–1721) had shown that no seed would grow without first being pollinated. Then, in 1717, the French botanist Sebastien Vaillant had lectured on the sexuality of plants, using the pistachio plant in the Jardin du Roi in Paris to make his point. Linnaeus took up this idea.

In 1729, Linnaeus wrote a paper called *Sponsalia Plantarum* in which he wrote about 'the betrothal of plants, in which ... the perfect analogy with animals is concluded'. Vaillant had talked about just the petals when talking about a flower's sexual organs. Linnaeus insisted that it is the stamens where pollen is made (the 'bridegrooms') and the pistils where seeds are made (the 'brides') that are the real sexual organs.

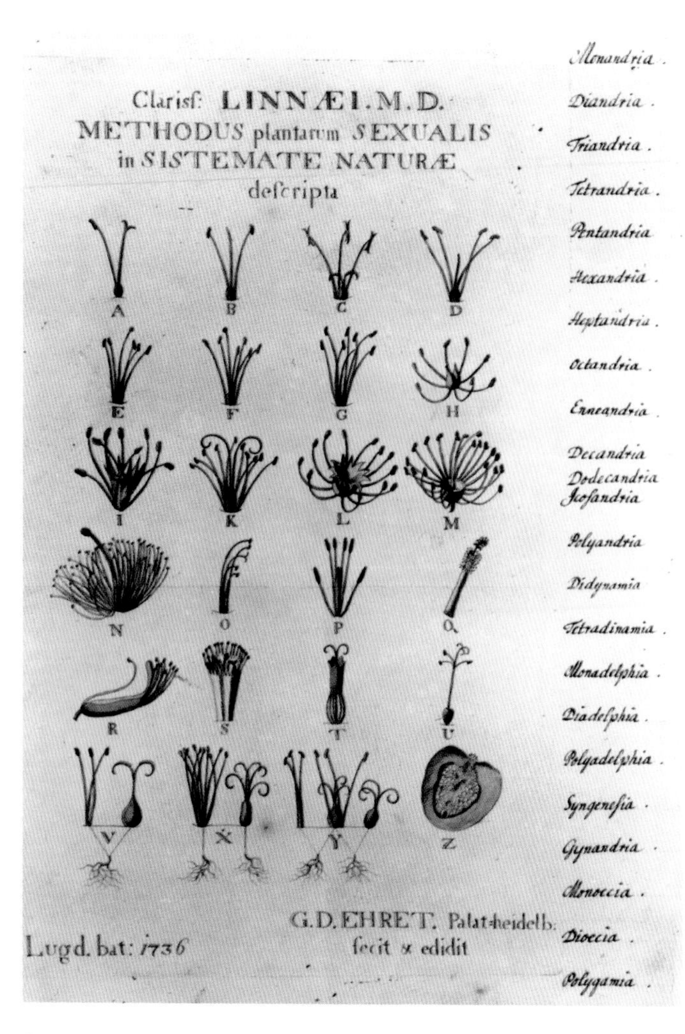

Ehret's original drawing illustrating Linnaeus's sexual system, from The Compleat Naturalist, *1736*

Brides and bridegrooms

Taking Ray's idea that species are basically living things that reproduce together, Linnaeus proceeded to develop a scheme based entirely on the sexual equipment of plants. He divided all flowering plants into twenty-three classes according to the length and number of their stamens – their male organs – then subdivided these into orders according to the number of pistils – female organs – they possessed. There was a twenty-fourth class, the *Cryptogamia*, that included plants like mosses, which appeared to have no flowers.

Many people were hugely offended by the sexual overtones in Linnaeus's scheme. One class he named *Diandria*, meaning 'two husbands in one marriage', while he said 'the calyx might be regarded as the *labia majora* or the foreskin; one could regard the corolla as the *labia minora*'. Indeed, some found it all so shocking that for almost a century botany was not seen by some people as a decent thing for young ladies to be interested in.

Linnaeus the teacher

Yet Linnaeus's scheme was simple and practical. Its great beauty was that anyone, with just a little training, could learn to identify which class a plant belongs to simply by counting its stamens. To prove his point and to train disciples, Linnaeus would lead enthusiastic plant-spotting hikes through the countryside around Uppsala when he returned there.

With groups of up to 300, Linnaeus would roam the fields and woods, gathering specimens before marching back into town accompanied by a band of musicians. Eventually, the Rector of the University put a stop to these hikes, feeling they were distracting students from their studies, saying, 'We Swedes are a serious and slow-witted people; we cannot, like others, unite the pleasurable and fun with the serious and useful.' But Linnaeus was never short of disciples.

The Swedish garden

Settled firmly in Uppsala, Linnaeus created his own botanic garden, laid out in the same order as his great classification. There he tried to plant specimens brought to him from around the world. Linnaeus believed that all the species of plant had existed in God's original Garden of Eden and had only become scattered since the Fall, so it was a pious act to bring them altogether.

Linnaeus also believed that Sweden could cater entirely for its own needs if every economic plant was grown within its borders. Oddly, for such an experienced botanist, he seemed convinced that even tropical plants could survive in Sweden's cool climate. When tender plants perished, he apparently just assumed they were weak specimens.

Throughout the 1740s, Linnaeus worked incredibly hard, cataloguing plant species and fitting them into his scheme. In 1745, he published a complete encyclopedia of Swedish plants called *Flora Suecica*. As he was working, he began to think about the naming of species.

The two-name system

His solution did not come easily. He wanted names that were accurate and complete but simple enough for the amateur to use on field trips. He felt amateur botanists could remember the genus a species belonged to, and then perhaps one more element to identify the species when they went home and looked it up. This gave him the idea of the two-name or 'binomial' system.

He was tempted to make the second name quite complex and fully descriptive of the species, but then he realized he only needed an easy-to-remember label for later reference. He resisted this at first, calling such names 'trivial'. Then, in 1751, he resumed work on his great project to catalogue all the world's plants and made his decision to put in the trivial second name, which was, he said, 'like putting the clapper in the bell'.

Realizing that he had to get the names in place before someone else gave plants other names, he gave a binomial label to every known species. In an incredible burst of invention, Linnaeus succeeded in naming 5,900 plant species in little more than a year, and in 1753 he published them all in his *Species Planetarium*.

With his work on the plant kingdom complete, Linnaeus turned his attention to the animal kingdom. In his earlier *Systema Naturae* of 1735, he had used the classification Quadrapeds (four-legged creatures), but now he realized that a more crucial characteristic than four legs was the presence of mammary glands for suckling young. So Quadrapeds were replaced by Mammals, and the first or prime group in the Mammals was the primates, which included humans, named *Homo sapiens* (Wise man) by Linnaeus. In 1758, Linnaeus published his catalogue of animals with binomial names in his tenth edition of the *Systema Naturae*.

By now, Linnaeus had many students, who would voyage all over the world to bring him his samples, while Linnaeus would sit at home waiting for their return like an anxious parent. In 1755 he had turned down an offer from the king of Spain to come and live at the Spanish court with a very handsome salary. In 1761, he was granted a royal patent to make him part of the Swedish nobility and he changed his name to Carl von Linné. Three years later, a stroke left him badly weakened, and he died on 10 January 1778.

Linnaeus's apostles

Throughout the 1740s and 1750s, many of Linnaeus's best students travelled around the world to investigate and bring back plants from distant lands. These 'apostles', as Linnaeus called them, were all young, ardent enthusiasts of Linnaeus's botany – 'true discoverers ... as comets among the stars', Linnaeus exclaimed proudly.

Their mission was often dangerous and five of them – Anders Berlin, Pehr Forsskål, Fredrik Hasselqvist, Pehr Löfling and Christopher Tärnström – never returned. Linnaeus often reproached himself for sending them on their way, but they were all keen to go and the plants they brought back or described hugely enriched Linnaeus's collection.

Pehr Kalm's trip to North America provided Linnaeus with 90 species of North American plants, 60 of them entirely new. Peter Osbeck brought 600 specimens back from China. Carl Thunberg described over 3,000 species in Japan, over 1,000 of which were entirely unknown. Solander joined Captain Cook on the *Endeavour* on its round-the-world voyage and gave names to 1,200 new species and 100 new genera of plants, plus a host of animals.

By the time Linnaeus died, it was the norm for expeditions around the world to take a botanist with them, culminating, of course, in Charles Darwin's famous voyage on the *Beagle*, and the riches they brought back increased scientific knowledge of the natural world immeasurably.

James Hutton

Scotsman James Hutton (1726–97) was the brilliant eighteenth-century geologist who finally revealed the ancient nature of the Earth and the long and gradual processes which shape its landscapes.

1726–97

IT SEEMS SO OBVIOUS TO US NOW THAT THE WORLD IS VERY OLD and has been shaped by geologic processes over millions of years that it is hard to believe that people did not always see it that way. Even as recently as the eighteenth century, in the Age of Enlightenment, most people still believed the Earth was not much older than human history. For many people, the truth about the Earth's age lay not in rocks and the landscape, but in the Bible. In 1650, the Irish Archbishop James Ussher gave the official line. After studying the Bible, he concluded that the world began on Sunday, 23 October 4004BC and has changed little since, except perhaps during the time of the Deluge, the biblical flood, which he dated to 2349BC. And the great Isaac Newton agreed!

Yet just as Copernicus's revelation that the Earth is not at the centre of the universe was finally sinking in, so some thinkers were beginning to question this view of the world. They could see that many rocks in the landscape were formed from sediments and that they were full of fossils of sea creatures, yet they had no idea how the sediments came to form mountains or how the fossils got there.

The leading idea in the eighteenth century was that it was all the result of a single great catastrophe that shaped the world quickly in one fell swoop and then left the landscape as it is. The great German geologist Abraham Gottlieb Werner proposed the theory that this catastrophe was a flood. Most rocks, he suggested, formed in a universal ocean that covered the entire Earth, and were then left behind as the landscapes we see today when the ocean waters drained away. Of course, many saw the biblical Deluge in Werner's universal ocean.

Hutton's great breakthrough was to show that this idea was wrong in two major respects. First of all, he realized that it was not flood waters that had formed many rocks and built mountains, but the Earth's internal heat – the heat of the molten rock that emerges in volcanoes. Secondly, he showed that landscapes were not shaped once and for all in some great catastrophe, but slowly and continually by countless cycles of erosion, sedimentation and uplift that are repeated over incredibly long time periods. Again and again, he realized, rocks are worn away, their debris is washed into the oceans to settle as sediments and the sediments are then uplifted and distorted by the heat of the Earth's interior to form new landscapes. If this is so, the Earth must be very very old – not just thousands of years, but millions. Hutton never stated how old he believed the world to be, but the implication was clear that it was extremely old, as others soon began to realize.

Scots upbringing

James Hutton was born in Edinburgh in June 1726, the eldest son of Sarah and William Hutton. William died just 2 years later. Edinburgh was going through an extraordinary time, with the last of the Scottish Highland rebellions turning the city into something of a frontier town. Hutton's childhood, though, was peaceful enough.

He was educated first at Edinburgh High School, then as a teenager at the University of Edinburgh, where he came under the eye of Colin Maclaurin, who had worked with the aging Newton in London and was much admired by him. Maclaurin introduced Hutton to Newton's ideas – notably the cycles of the planets – which had a marked influence on him. Maclaurin also introduced Hutton to the idea of deism – the idea of a God who designed and created the universe as a perfect machine, then left it to run by itself. This idea played a big part in Hutton's way of thinking – and helped him to see that the idea of an ancient Earth did not conflict with belief in God.

There was nothing in James Hutton's life at this time to indicate his later interest in the Earth. When he left university in 1745, he went to medical school to train as a doctor. It was the same year that Charles Stuart, Bonnie Prince Charlie, led the last Scottish rebellion. No doubt the well-to-do in Edinburgh did not approve of the brutality with which the Highlanders were treated in the aftermath of Charlie's flight into exile, but it set the city free to flourish in an astonishing way. Elegant new streets were laid out and the beginnings of an intellectual golden age began that was to see Edinburgh dubbed the 'Athens of the North'.

The layers of rock in this unconformity are seen as ridges running along the contours of the eroded mountain.

Doctor Hutton, farmer

Yet even as the echoes of the rebellion were still ringing in the streets, the young Hutton was obliged to leave the city in embarrassment, having made a young servant girl pregnant. He went to Paris and studied medicine there for five years, before completing his medical training at Leyden in Holland.

Strangely, though, he never seemed for a moment to have contemplated becoming a doctor. In 1750, he went into business with an old Edinburgh friend to make sal ammoniac, a key ingredient in steel-making, and this was to provide him with a moderate income all his life. Then, in the 1750s, he inherited the family farm at Slighhouses south of Edinburgh, and he decided to become a farmer. Before he did so, he set out to learn all about the latest farming techniques, and he soon turned Slighhouses into one of the most innovative, prosperous farms in Scotland.

Hutton's interest in the Earth seems to have begun while he was learning about farming. He travelled extensively around Britain, studying the landscape and rocks, bringing sample after sample of rocks and minerals back home – something quite new in those days – and he soon seems to have become known for his expertise on them. His close involvement with the land had a crucial influence on his geological ideas. Watching the land on his farm change year by year, with soil washed away by winter rains only to be replenished through the years as rock was weathered, may well have been the inspiration for Hutton's vision of endless cycles of erosion and renewal. By the early 1660s, his ideas were already partly formed.

Return to Edinburgh

With the farm well-established, Hutton moved back to Edinburgh in 1770 and was immediately caught up in the intellectual ferment gripping the city. Here he met and befriended, among many other great minds: David Hume, the great philosopher of human nature; Adam Smith, whose book *The Wealth of Nations* became the bible for the rise of capitalism; James Watt, who made the steam engine practical; and Joseph Black, who discovered carbon dioxide. Black and Watt came to be among Hutton's closest friends and supporters.

Hutton realized that it was heat from within the Earth that was involved in mineralization – heat and extreme pressures from deep in the Earth.

Before long, Hutton and Black teamed up to work on what was one of the key problems in geology at the time – just how did all the different minerals of which rocks are made form? Most mineralogists of the day believed, like Werner, that they precipitated out of the universal ocean. But if this was so, then surely every substance found in rocks must dissolve in water. This clearly was not the case. During his work with Black, Hutton realized that it was instead heat from within the Earth that was involved in mineralization – and not just heat, but heat and extreme pressures such as are only found deep within the Earth.

Going public

Hutton's ideas began to crystallize and in 1785 he decided to present them to the Royal Society of Edinburgh. Perhaps suffering from stage fright, he left Black to read his paper. Some people were struck at once by the significance of his ideas, while others may have objected that there was no proof. Either way, Hutton was determined to find all the proof needed to convince doubters, and set off with his friend Sir John Clerk into the Highlands to find it.

In particular, Hutton was interested in granite rock, an 'igneous' rock which forms from molten magma from inside the Earth. Hutton wanted to show that outcrops of granite may often form after, not before, the sediments surrounding it. If so, Werner's idea could not be true that all sediments were formed by the universal ocean and were simply laid down on top of primitive igneous rock. In 1788, to Hutton's excitement, they saw the graphic proof he was searching for in Glen Tilt in the Grampian Mountains, where granite veins had clearly been injected into the surrounding rock. Later that summer, Hutton took John Playfair and James Hall to Siccar Point on the North Sea coast to show them the unconformity (see box) and he had all the proof he needed.

Convincing the world

But if Playfair and Hall were convinced, there were many who weren't. Hutton could deal with those who toed the established scientific line and criticized him for lack of understanding; he knew his evidence was sound, and he had many supporters in Edinburgh to back him up. But in 1793, a leading Irish academic called Richard Kirwan suggested Hutton's theories were blasphemous.

Hutton was determined to respond, but that year he fell seriously ill, probably with kidney failure. He began to write a book called *Theory of the Earth*, explaining his theories and giving the proofs. Unfortunately, he was too ill to make a good job of it, and he was unable even to write the last of the three volumes, containing all his proofs. When Hutton died in 1797, his messy book, which was much less clear than his earlier papers, made little impact.

Playfair and Hall took up his cause, Playfair writing a simple summary of his ideas, Hall conducting laboratory experiments to show that igneous rock could form mineral crystals simply by cooling down slowly. All the same, it was to be another 35 years before Charles Lyell wrote his famous book *Principles of Geology*, which made Hutton's ideas the foundation of modern geology and provided Charles Darwin with the inspiration for his theory of evolution.

Unconformity

A key element in the proof of Hutton's theories was the existence of unconformities – places where there is a clear break in the normally orderly pattern in which rock layers are laid down over time, one on top of another. In the summer of 1788 Hutton took two potential critics, John Playfair and James Hall, along the North Sea coast to Siccar Point in a rowing boat. Playfair later described the experience: 'On us who saw these phenomena for the first time, the impression will not easily be forgotten ... We often said to ourselves, What clearer evidence could we have had of the different formation of these rocks, and the long interval which separated their formation, had we actually seen them emerging from the bosom of the deep.'

What they saw that afternoon was an unconformity. There in the cliff face were layer upon layer of rock, not lying horizontally, but standing straight up, like books on a shelf. Then, on top of them, cutting across them almost horizontally, were more layers of rock, this time lying almost flat. This cutting across is the unconformity. It was clear, as Hutton explained to his companions, that the vertical layers were originally laid down flat as sediments, then uplifted and tilted almost upright. Erosion had cut across the top of these layers, creating a new plain or seabed, and then new sediments had gradually settled on this and built up the horizontal layers of sediment, which had in turn been uplifted to create the top of the cliff. No more convincing proof of Hutton's idea of endlessly repeated cycles of erosion, sedimentation and uplift could be seen, and Playfair and Hall instantly became Hutton's most ardent champions.

Antoine Lavoisier

Antoine Lavoisier (1743–94) is said to be the father of chemistry, making the first list of elements, establishing the idea of conservation of mass and discovering the true nature of burning and the role of oxygen.

1743–94

Thanks to Newton and Galileo, scientists in the eighteenth century knew a lot about how and why things moved, but little about what things are made of. Chemistry was still closely linked with alchemists searching for what they called the philosophers' stone, which would turn ordinary 'base' metal to gold. The idea of chemical elements was still in its infancy, and most scientists still believed, as the Greeks had done, that there were just four elements – air, water, earth and fire.

At the beginning of the century, alchemists were in some ways still leading the field in the study of matter. It was a German alchemist, George Stahl, for instance, who came up with a theory to explain how things burn – the phlogiston theory, which was to dominate scientific debate for half a century. Stahl suggested that anything burnable contains a special 'active' substance called phlogiston that dissolves into the air when it burns.

Gradually the debate over the nature of matter began to shift towards the chemists and a more down-to-earth approach demanding experimentation, close observation and proofs. In part this came from Britain, where the impetus of the Industrial Revolution prompted brilliant but highly practical men like Joseph Priestley to try their hands at chemical experiments. But the great pioneer was undoubtedly a Frenchman – Lavoisier.

Lavoisier benefited from the work of British chemists, particularly Priestley, and there have been some arguments over who achieved what. But it was Lavoisier who pulled it all together and made many significant advances in his own right.

It was Lavoisier, for instance, who realized that every substance can exist in three states or phases – solid, liquid and gas. And, by suggesting that a gas is a substance, he opened the way to the idea that air not only has mass but may be a mixture of gases. It was Lavoisier who finally showed that air is a mix of two main gases, oxygen and nitrogen (which he called azote). It was Lavoisier, too, who, with others, showed that water is a compound of two gases, hydrogen and oxygen. And it was Lavoisier who finally proved that the phlogiston idea of burning was wrong and gave us our modern theory of combustion.

Lavoiser was a meticulous experimenter who championed the notion of exact measurement and the idea of conservation of mass, which is that no matter how substances change in an experiment, no mass is ever lost. This crucial insight not only helped him prove the true nature of combustion (see box on p. 75), but still underpins all experiments with matter even today. In his famous *Traité Élémentaire de Chimie* (*Treatise on Chemical Elements*), in which he explains clearly for the first time how chemical compounds are formed from elements, he wrote, 'We must trust to nothing but facts. These are presented to us by Nature and cannot deceive. We ought in every instance to submit our reasoning to the test of experiment ...'

The young lawyer

The eldest son of a lawyer, Jean-Antoine Lavoisier, and Émilie Punctis, Antoine Lavoisier was born in Paris on 26 August 1743. His mother died when he was 3 and he was then brought up by his adoring aunt Clémence Punctis. The Punctis family were very wealthy, and so young Antoine was brought up in comfort and went to the elite Collège Mazarin, where he studied both science and law. His legal training undoubtedly helped him present his arguments clearly, but also alerted him to the need for proofs. In his spare time he attended lectures on chemistry.

When Lavoisier graduated as a lawyer in 1763, he went with a friend of the family, geologist Jean-Étienne Guettard, on a trip through France cataloguing minerals. Then, in 1765, using his newly acquired chemical and mineralogical knowledge, he presented a report to the Academy of Sciences in Paris on the nature of gypsum, which is used to make plaster of Paris. A common thread through Lavoisier's life was his interest in public works, and when he was admitted to the Academy the following year at the young age of 23, it was partly for a brilliant paper on the way to light the streets in Paris.

It was about this time that he received a large inheritance from his mother, and he used it to buy an interest in a company called the General Farm. The Farm was a company that paid the government money, but in return was allowed to collect taxes. It was a sound financial investment, and it helped to make Lavoisier wealthy. But it would also prove in time to be his downfall, for the tax farmers were very unpopular with the people.

The Farm also brought Lavoisier his wife, Marie-Anne-Pierrette Paulze, the pretty 13-year-old daughter of another tax farmer. Despite the age difference – Lavoisier was almost 30 when they married – it proved to be a happy marriage. Marie-Anne proved a more than able scientific assistant, not just helping Lavoisier in his laboratory, but taking drawing lessons from the famous painter Jacques-Louis David so that she could illustrate his work, and learning English in order to translate the work of the British scientists for him.

'We must trust to nothing but facts.... We ought in every instance to submit our reasoning to the test of experiment.'

Lavoisier in the laboratory

Lavoisier's wealth meant he could build himself what was perhaps the best-equipped chemistry laboratory of the age, and he began to undertake a series of important experiments.

First of all, he began to try out the phlogiston theory. After testing burning sulphur, phosphorus and other chemicals, he suggested a new theory – that when things burn they do not give off phlogiston but take in air. This wasn't the whole truth, but it was a significant step, and he decided to investigate the findings of the British scientists with the various substances in air. His idea was, as he said, 'to link our knowledge of the air that goes into combustion or is liberated from substances, with other acquired knowledge, and to form a theory' – and his efforts were successful, bringing him to what is essentially the now accepted theory of combustion (see box).

Air and water

In 1774, the English chemist Joseph Priestley was watching what strong heat could do to mercuric oxide. He noticed that it gave off a gas and that, to his surprise, a candle would burn in this gas with an unusually strong flame. On a visit to Paris the following year, he met Lavoisier and told him about this phenomenon. Lavoisier at once did a series of experiments with the new gas and with air.

Lavoisier learned from these that air is made up from two gases: firstly, Priestley's gas, the gas involved in combustion, which Lavoisier called oxygen, and secondly the gas that came to be called nitrogen, and which he called azote. Although this is now forgotten, Lavoisier wrongly coined the word oxygen from the Greek for 'acid-forming' which oxygen isn't.

Even more importantly, Lavoisier went on to show that burning is closely related to breathing, and that both involve oxygen. Our lungs take in the oxygen we need from the air and expel carbon dioxide. He also showed that oxygen reacts with metals to form oxides, a process called oxidation. Rusting, rotting organic matter and burning wood are all kinds of oxidation.

Not content, Lavoisier went on to experiment with water. Priestley and other British scientists, such as Henry Cavendish, had already noted that oxygen and hydrogen could be changed into dewdrops when an electric spark brought them together. Lavoisier identified the dew as water and showed that it was created by joining hydrogen and oxygen.

Lavoisier was aware how important his work was, and was determined to show that he was establishing a new field of science, experimental chemistry. Firstly, in 1787, he published a method for naming chemicals according to their properties,

A coloured illustration of an original drawing of Lavoisier's laboratory by his wife and assistant Marie-Anne.

Phlogiston and burning

The phlogiston debate was quite literally the burning scientific question of the eighteenth century. It was the focus of the battle between old ideas, which owed much to alchemy, and the new science of chemistry – and in particular the idea of elements.

The alchemists still believed in the same four basic elements as the ancient Greeks – air, water, fire and earth, while the chemists leaned to the ideas of Robert Boyle. Boyle had suggested the idea of chemical elements – that everything is made from a handful of basic substances or 'elements', each made from a tiny lump called an elementary corpuscle. If Boyle was right, then elements could only be mixed together, not changed – and the idea of the four elements might have to be abandoned.

The debate between alchemists and chemists became centred on the phlogiston theory, which said that anything burnable contained a special 'active' substance called phlogiston that dissolved into the air when it burned. Therefore anything that burned must become lighter because it loses phlogiston. At first this theory had seemed sound to everybody and had become established as scientific orthodoxy. Later it began to be challenged, and it was Lavoisier who realized that the way to test if it was true was to weigh substances carefully before and after burning. In a famous and brilliant experiment, he burned a piece of tin inside a sealed container – and found that, contrary to phlogiston theory, the tin actually became heavier after burning, while the air became lighter. There was no change in mass at all, as the substances were simply changing places. It was also clear that rather than losing something (phlogiston) to the air, the tin was taking something from it. Later, Lavoisier realized that this something was the gas he called oxygen.

Lavoisier's experiment was a turning point in our understanding of matter for three reasons. Firstly, it put accurate scientific measurement firmly at the heart of chemistry. Secondly, it demolished the phlogiston theory and showed that burning is a process involving oxygen. Thirdly, it showed that substances do not change or vanish even in as dramatic a process as burning; they simply swap places.

and the system of symbols as a shorthand for them that chemists still use today. Secondly, he wrote a major summary of the field in his *Traité Élémentaire de Chimie* of 1789.

Social science and revolution

While his laboratory work was important, Lavoisier was also very concerned with more practical projects. Throughout the 1770s and 80s, Lavoisier conducted a whole series of studies and compiled over 200 separate reports on a huge range of issues, including food adulteration, how dyes work, how water could be stored aboard ship, how glass could be made better, how prisons could be improved, whether canal water could be drunk and much more besides. He approached each of these with the same thoroughness and insight that he did with laboratory work, and made many genuine improvements in social conditions.

All this, however, could not help him when the Revolution came. In the terror of 1793, Jean-Paul Marat came to power, and the revolutionaries decided it was time to pay off old scores – particularly against those who had profited from the old régime as the unpopular tax farmers had done. Lavoisier was tried and found guilty, and when his achievements in science were brought to the attention of the judge in an attempt by his friends to save him, the judge is said to have replied, 'The Republic has no need of scientists.' Lavoisier was guillotined on 8 May 1794. A witness to the event, the famous mathematician Joseph Louis de Lagrange, commented, 'It took a mere instant to cut off that head, and yet another hundred years may not produce another like it.'

John Dalton

A mild-mannered chemist who is sometimes said to be the father of chemistry, John Dalton (1766–1844) established the modern theory of atoms and elements, and so paved the way for a huge range of scientific achievements.

1766–1844

THE IDEA OF ATOMS WAS BY NO MEANS NEW in the eighteenth century. In fact, it had been around for well over 2,000 years. For instance, the great Greek thinker Aristotle (384–322BC) thought that in theory matter could be chopped up into smaller and smaller pieces, but other Greeks, notably Democritus (460–400BC), argued that matter was really made of tiny particles with empty space in between, just as scientists believe today. These particles were the smallest possible pieces of matter, which is why Democritus called them 'atoms', from the Greek for uncuttable.

People struggled to imagine how air or water could be made of tiny particles, and although the idea of atoms was not forgotten, Aristotle's view was found to be more convincing. In addition, Aristotle's view that the world was made from just four main elements – earth, water, air and fire – also proved convincing. Even now, scientists believe matter exists in four phases – solid, liquid, gas and plasma – which in some ways correspond to Aristotle's four elements.

Then, in the seventeenth century, scientists began to question Aristotle's view of matter. The Irish scientist Robert Boyle (1627–91) suggested that there were other kinds of element – basic pure substances that could combine to make compounds. Crucially, he suggested that each of these 'chemical' elements had its own unique characteristics and that each could exist as a solid, liquid or gas. Boyle even suggested that matter might consist of atoms after all.

The Aristotelian view of matter was further undermined as experiments began to reveal that neither air nor water were indivisible elements. First of all, such scientists as Joseph Priestley (1733–1804) and Antoine Lavoisier (1743–94) showed that air is a mixture of gases, including oxygen and nitrogen. Then Lavoisier showed that water, too, is a compound, of hydrogen and oxygen. Lavoisier even made a list of a dozen basic chemical elements, including these newly discovered components of air and water. After over 2,000 years, the atomic idea was finally beginning to gain credence.

Yet for all this progress, no one knew just what an element was – and no one had thought to connect them with atoms in any way. It was assumed, for instance, that if matter, including air, was made of atoms, all atoms must be identical. It was John Dalton's great insight to see that the atoms for each of the gases in air might be different. He then went further to unify all the progress of the previous century in the atomic theory of elements which underpins science today. Dalton suggested that all the atoms of an element are identical – but different from every other element. He also argued that compounds were formed by the joining of an atom of one element with an atom of another. Although theories of elements and compounds have developed since Dalton's time, the essentials of his ideas are all still there.

Childhood in the Lakes

John Dalton was born on 5 or 6 September 1766 in the English Lake District coastal town of Cockermouth to a family of Quaker tradesmen. As a boy he was alert and inquisitive rather than a brilliant scholar. But he was bright enough to be appointed as teacher at his local school at the age of 12, and a Quaker relative, Elihu Robinson, tutored him in science.

In 1781, at the age of 15, he went across to Kendal on the far side of the Lake District to teach at a boarding school. While there he was introduced to mathematics and natural sciences by the remarkable blind philosopher John Gough, who is described by the poet William Wordsworth in his poem *The Excursion*: 'Methinks I see him now, his eyeballs roll'd beneath his ample brow.' Encouraged by Gough, and by the often awe-inspiring Lake District weather, Dalton began to make meteorological observations.

Weather man

Over the next 15 years, Dalton made a weather diary in which he recorded over 200,000 observations. His interest in the weather was lifelong and profound, and his insights groundbreaking. Indeed some people believe he ought to be called the 'father of meteorology' as well as the 'father of chemistry'.

John Dalton collecting marsh gas. It was his observation of gases that first planted the seeds of atomic theory in Dalton's mind.

Dalton was always wary of trusting other people's ideas, and insisted that he would believe in only what he observed for himself. 'Having been in my progress so often misled by taking for granted the results of others,' he wrote, 'I have determined to write as little as possible but what I can attest by my own experience.' He was not a particularly accurate observer, nor a brilliant experimental scientist, but his sceptism, plus a remarkable capacity for insightful overview, enabled him to make the most of his work.

Over the years, he wrote important papers on the barometer, the thermometer, the hygrometer, rainfall, the formation of clouds, evaporation, the movement of atmospheric moisture, and much more. Dalton was the first to realize that atmospheric moisture turns to rainfall not as a result of pressure changes, but because of reductions in temperature, reducing the air's capacity to hold water vapour. In 1788, after witnessing a spectacular display of the aurora borealis, or Northern Lights, he came to the remarkably prescient conclusion that auroras were caused by the Earth's magnetism. Dalton – unaware that George Hadley had already suggested the same thing – even came to the correct conclusion that the trade winds were caused by a combination of regional temperature variations and the Earth's rotation.

Colour blindness

Besides his meteorological studies, Dalton also became fascinated by a condition that he and his brother shared – colour blindness. Dalton was the first to study colour blindness scientifically, and the condition is still sometimes known as Daltonism. Colour blindness was the subject of his first paper, entitled 'Extraordinary facts relating to the vision of colours', that he presented when he was elected to the Manchester Literary and Philosophical Society in 1791.

Dalton asked that when he died, his eyes be examined to discover the reason for his colour blindness, because he believed the fault could be that the fluid in his eyes was blue. A postmortem showed the fluid was entirely normal. But in the 1990s, DNA tests on his eyes, which have been preserved at the Royal Institutution for 150 years, showed they lacked the pigment needed to give sensitivity to green.

A world of particles

Dalton contributed hundreds more papers to the Manchester Society and in 1817 was elected its president, a post he held for the rest of his life. Perhaps the most important of these papers were presented in the early 1800s. By that time, Dalton had given up full-time teaching and was funding his scientific studies through private tutoring of the sons and daughters of Manchester's rising middle class.

The extra time this gave him enabled him to produce this series of papers, in which he developed his atomic theory of matter (see box on p. 79). He argued that the elements then known, including hydrogen, oxygen and nitrogen, are made up of atoms – that is, 'solid, massy, hard impenetrable, moveable particles'.

It was the key insight that chemists had been looking for – an understanding of just what elements were, and why they combined as they did. The implications of the theory were accepted surprisingly rapidly, and quickly became the focus of most chemical research.

Dalton's atomic theory

The earliest of the papers in which Dalton developed his atomic theory concerned his studies of air pressure and the way it affected how much water could be absorbed. Contrary to the prevailing view of the day, Dalton was convinced that air is a mixture of different gases, not a compound of them like water. In his experiments, he observed that pure oxygen will not absorb as much water vapour as pure nitrogen – and jumped to the remarkable intuitive conclusion that it was because oxygen atoms were bigger and heavier than nitrogen atoms. 'Why does not water admit its bulk of every kind of gas alike?' Dalton wrote, 'This question I have duly considered, and though I am not able to satisfy myself completely, I am nearly persuaded that the circumstance depends on the weight and number of the ultimate particles of the several gases.'

In a paper read to the Manchester Society on 21 October 1803, Dalton went further, saying, 'An inquiry into the relative weight of the ultimate particles of bodies is a subject, as far as I know, entirely new; I have lately been prosecuting this enquiry with remarkable success.' He went on to describe how he had arrived at different weights for the basic units of each elemental gas – in other words the weight of their atoms, or atomic weight. He went on to argue that the atoms of each element combined to make compounds in very simple ratios, and so the weight of each atom could be worked out by the weight of each element involved in a compound, an idea which later came to be called the Law of Multiple Proportions.

Hydrogen, Dalton realized, was the lightest gas, and so he assigned it an atomic weight of 1. Because of the weight of oxygen that combined with hydrogen in water, he assigned oxygen an atomic weight of 7. There was a basic flaw in Dalton's method, because he did not realize that atoms of the same element can combine. He always assumed that a compound of atoms, a molecule, had only one atom of each element. It was another half century before scientists realized that only the Italian scientist Amadeo Avogadro's idea of using molecular proportions would allow them to calculate atomic weights correctly. Nevertheless, the basic idea of Dalton's atomic theory – that each element has its own unique-sized atoms – has proved to be resoundingly true.

Some scientists have played down Dalton's role, suggesting that his scientific research was less than solid, and that his atomic theory was built on the work of others. However, there is no doubt that it was Dalton who brought it all together and into mainstream science – and there is no doubting the importance of atomic theory to science. As his biographer Frank Greenaway says, with Dalton's theory 'we have made new materials, utilized new sources of energy, defeated one disease after another, and come within sight of the mechanism of life'.

The unworldly hero

Dalton's work as a teacher had already made him much loved. With the establishment of his atomic theory, his scientific reputation became immense, yet he remained a solitary man, never marrying, rarely socializing and almost living the life of a recluse. His time was dedicated to his pupils and to his scientific research. Throughout his life his needs were simple and his habits plain, and he always dressed in the plain Quaker style.

In 1810, his scientific achievements were recognized with an invitation to join the Royal Society. Diffident and short of money, Dalton declined the offer, but then in 1822 the Royal Society elected him a member and paid his election fee for him. The French Academy of Sciences made him their sole permitted foreign member.

Dalton lived out his last years quietly. His scientific researches began to look rather introverted, as, in his dogged independence, he refused to acknowledge the advances made in science – and valuable corrections to his ideas – in the years since he had announced his atomic theory. Nonetheless the respect and affection in which he was held became clear when he died of a stroke at the age of 78. Tributes poured into the Manchester Society, and the people of Manchester gave him what can only be called a state funeral, attended by 40,000 mourners.

Michael Faraday

Michael Faraday (1791–1867) was not merely a great experimental scientist – perhaps the greatest of all time – but a visionary theorist who realized that all the forces of nature are interconnected. Faraday gave us both the electric motor and the electric generator, paving the way for the electrical revolution.

1791–1867

When Michael Faraday was born in 1791, electricity was the new hot topic in fashionable society. Scientists and showmen of the day were able to create dramatic sparks by turning a wheel to rub glass against sulphur to generate static electricity. An Italian anatomist Luigi Galvani (1737–98) was using electricity to make the legs of dead frogs twitch. Galvani believed he had discovered the very force of life, 'animal electricity', that animated flesh and bone. Soon dozens of scientists were trying to bring corpses back to life by electrifying them, a spectacle all too well captured in Mary Shelley's *Frankenstein*.

This idea of electricity as a life force gained a powerful hold on the public imagination, and inspired a host of fanatical scientists, including Andrew Ure, who, in a gruesome display in 1818, made the corpse of executed Glasgow murderer Matthew Clydesdale dance like a puppet. Another scientist, Andrew Crosse, in 1836, claimed to have created insects called acari with electricity. But while all this electrical hysteria was going on, rapid and serious advances were being made by experimental scientists towards understanding the true nature of electricity.

In the late 1790s, for instance, Alessandro Volta realized that electricity can be created by a chemical reaction, and he used this idea to create the first battery in 1800. Using Volta's battery, or 'trough', to give them a supply of electricity on tap, various scientists discovered that electricity would flow through a complete loop or circuit. André Ampere also learned about the strength of currents, and Georg Ohm discovered the nature of electrical 'resistance'.

Then, in 1820, the Danish scientist Hans Oersted discovered that an electric current could make the needle of a magnetic compass swivel. It was the first inkling of a link between two natural forces, and it immediately became the focus of experimentation among scientists. Faraday was just one among many who tried their hand at unlocking the secrets of electricity and magnetism in the 1820s and 1830s.

His extraordinarily inventive and meticulous experimental work and his keen theoretical insight, however, put him at the forefront of all the key breakthroughs. Within a few months of hearing of Oersted's discovery, Faraday had constructed an ingenious demonstration that showed how a magnet would move in a circle around an electric wire, and an electric wire would move in a circle around a magnet. He had discovered the principle of the electric motor.

Ten years later, Faraday made an even more important discovery, that moving a magnetic field can create or 'induce' a current of electricity. This principle of electromagnetic induction, which was discovered independently by Joseph Henry in America around the same time, meant that machines could be built to generate huge quantities of electricity, opening the way for everything from electric lighting to telecommunications.

Yet neither the electric motor nor the principle of electric induction are perhaps Faraday's greatest achievements. He went on not just to demonstrate the principle of electrolysis – the way chemicals are broken down by electricity – but to demonstrate the ultimate unity between all forces, including electricity, magnetism, light and even gravity, and to develop the idea of fields of force (see box on p. 83). This crucial insight has paved the way for all of modern physics and a host of modern technologies, from television to mobile phones.

Poor boy

When Faraday was born, science was very much the province of the rich. Not only was money needed to conduct experiments, but connections were also required to get on in the scientific establishment. Yet Faraday was the child of poor parents, and his story is often presented as a rags-to-riches tale. The British prime minister Margaret Thatcher saw him as the archetypal self-made man, triumphing over his poor background.

Faraday was brought up in a poor district of London near the Elephant and Castle. His father was a blacksmith, but so disabled that he was barely able to support his family. Faraday recalled that they often had only bread to eat for weeks at a time. He considered himself lucky, therefore, to be taken on as an apprentice errand boy in George Riebau's bookshop in Blandford Street at the age of 13.

At the bookshop, he didn't just run errands, but began to learn how to bind books, and also began to read them avidly – especially those on science. 'In early life,' he later wrote, 'I was a very lively imaginative person, who could believe in the "Arabian Nights" as easily as the "Encyclopedia", but facts were important to me, and saved me. I could trust a fact.' He became fascinated by science, and his helpful employer allowed him to set up his own makeshift laboratory in the bookbindery.

The magician's apprentice

By a stroke of good fortune, William Dance, one of Riebau's customers, was a member of the Royal Institution, the premiere scientific establishment of the day. Hearing of young Faraday's interest in science, Dance gave Faraday tickets to the celebrated demonstrations at the Institution of the famous scientist Sir Humphry Davy. It was the chance of a lifetime for a poor boy, and he took it with both hands. Faraday took avid notes at the lectures, drew them up meticulously, with illustrations, then sent a bound copy of them to Davy himself.

A museum worker demonstrates a Faraday cage. The cage is designed so that no electrical discharges will reach inside, and is used to protect electrical equipment from lightning strikes.

Suitably impressed, Davy took on Faraday, then 21, as his assistant at the Institution, and when Davy went on a tour of Europe the following year, he took Faraday with him, introducing him to many of the world's leading scientists, including Ampere, Volta and Gay-Lussac. Far from being self-taught, Faraday was getting his scientific education at the hands of the greatest minds of the age – and it paid off. Within a few years, he was not just helping Davy with his experiments, but actually conducting his own.

In 1821, the Royal Institution, in the wake of Oersted's discovery, asked Faraday to compile a survey of current research on electricity. Faraday not only did this with his now characteristic meticulousness and insight, but tried his own experiments, including the first demonstration of the principle of the electric motor. Davy was said to have been upset that his apprentice gave him no credit for this, since he had been working on the same problem himself, and some historians say it still rankled enough for him to vote against Faraday when he was invited to join the Royal Society – though it may be that Davy was simply seeing through his declared stance against nepotism.

'There is no more open door to study natural philosophy than by considering the physical phenomena of a candle.'

Faraday the presenter

Faraday was keenly aware of the need to present his discoveries to society, and so he hired the renowned teacher of public speaking, Benjamin Smart, to teach him how to present himself well to an audience. In 1826, he began his famous series of lectures for the public, which he called Friday Evening Discourses. He prepared these meticulously, with wonderfully effective demonstrations and experiments, and they became immensely popular, especially amongst the fashionable set. In the most spectacular of these he placed himself inside a steel cage while gigantic sparks of electricity were shot around the outside. He knew that the shape of the cage would protect him, but it was nonetheless a stunning display.

Even more successful were his Christmas Lectures for children, a tradition still practised at the Institution. The most famous of Faraday's Christmas Lecture series was called the Chemical History of a Candle, in which he used a candle to introduce a wealth of scientific ideas, from chemical elements to human respiration. 'There is no better,' he said, 'there is no more open door by which you can enter the study of natural philosophy than by considering the physical phenomena of a candle.'

By 1830, Faraday had become firmly established at the Royal Institution. His laboratory was in the basement. He lectured to the public on the ground and first floors. And he lived with his loving wife Sarah in a flat upstairs.

Reaching the peaks

By the time Faraday reached the age of 50, he was beginning to suffer frequent headaches and occasional memory loss. He was still immensely fit, able to complete a 73 km walk over the Alps in little more than 10 hours. But he was in more and more need of what he called 'head-rest'. Nevertheless, this was the time of what some regard as his greatest achievements.

In 1845, Faraday began a series of experiments in which he tried to find out if electromagnetism could affect the way light is polarized by transparent substances.

After experimenting with many substances, he finally tried a piece of heavy lead glass, and at once found that the polarization was affected by magnetism. It was an extraordinary achievement, showing the clear link between light and magnetism and electricity for the first time, which opened the way for the discovery of the complete spectrum of electromagnetic radiation, including television waves, microwaves, X-rays and gamma rays, as well as light.

It was about this time that Faraday began to talk about fields of force. The word 'field' actually came from William Thomson, the young Glasgow mathematician who – until the brilliant James Clerk Maxwell came along – Faraday felt was the only person in the world who fully understood his ideas. But the idea of fields was all Faraday's (see box).

The final years

Throughout the 1840s, Faraday kept more and more to himself. This was partly because of his religion. He was an ardent member of the small Sandeman sect, which was so strict about religious observance that they apparently suspended Faraday as an Elder when he missed a Sunday in order to accept an invitation from the queen. Interestingly, his religion meant that he could not accept all the honours he was offered, including a knighthood. People joked, 'Not so much Far-a-Day as Near-a-Knight.'

A quiet life was also imposed on him by his struggle against mental frailty. He relied increasingly on his wife Sarah to be a 'pillow for my mind'. He had ever more frequent dizzy spells, headaches and memory loss. In 1862, he wrote to his friend Schönbein, 'Again and again I tear up my letters, for I write nonsense. I cannot spell or write a line continuously. Whether I shall recover – this confusion – I do not know. I will not write any more.'

Faraday was given a Grace and Favour residence at Hampton Court Palace by the queen and died there on 25 August 1867 at the age of 76. He was buried in Highgate Cemetery in London.

Forces of nature

Newton had, with his concept of gravitation, made respectable the idea of an invisible force that exerted its effect through empty space, but this idea of 'action-at-a-distance' was beginning to look shaky to an increasing number of scientists in the early nineteenth century. By 1830, Thomas Young and Augustin Fresnel had shown that light did not travel as particles, as Newton had said, but as waves or vibrations. But if this was so, what was vibrating? To answer this, scientists came up with the idea of a weightless matter called 'ether'.

Faraday had another idea. He came to believe in the idea of fields made up of lines of force – the lines of force demonstrated so graphically by the patterns of iron filings around a magnet. This meant that action at a distance simply did not happen, but things moved only when they encountered these lines of force, which were not imaginary, but had a physical reality. Faraday appreciated that magnets induced electric currents by creating moving lines of magnetic force that carried an electrical charge as they moved.

The idea of fields of force is almost taken for granted, but in Faraday's time it was so radical that few even understood it, let alone agreed with it. They could see the idea of areas of magnetic influence, but the idea of electro-magnetic fields was completely beyond them. Mathematicians dismissed Faraday's ideas for their lack of mathematics. In 1855, Faraday wrote , 'How few understand the physical lines of force. They will not see them, yet all the researches on the subject tend to confirm the views I put forth many years since ... I am content to wait, convinced as I am of the truth of my views.' And he was right.

Charles Babbage

Sometimes called an 'irascible genius', Charles Babbage (1791–1871) was the remarkable English mathematician whose ideas for mechanical calculators and 'thinking' machines anticipated the computer age by 150 years.

1791–1871

CHARLES BABBAGE, SO THE STORY GOES, began his lifelong quest to create a mechanical calculating engine one evening in 1821. That night the young Babbage and his friend John Herschel were pouring over manuscripts of some mathematical tables they were preparing for the Astronomical Society, painstakingly checking the tens of thousands of entries one by one. As they did, they came across error after error made by the 'computers', the poorly paid human calculators who worked out such figures. Finally, in exasperation, Babbage exclaimed, 'I wish to God these calculations had been executed by steam!'

Babbage's frustration was not simply at the mind-numbingly tedious task of compiling tables, but the high chance of mistakes. Yet at the time such tables were vital in many spheres of life – science, taxation, engineering, surveying, insurance, banking, and more. When a ship set sail, for instance, the navigator's cabin was lined with volume after volume of tables to help him pinpoint the ship's position at sea.

People had been using aids to calculation for thousands of years – tally sticks, abacuses, and others – and through the seventeenth and eighteenth centuries, mathematicians such as Gottfried Liebniz and Blaise Pascal had created mechanical calculating aids. Some of these devices were exquisite and ingenious, but were limited in scope and prone to error, both because they could give a wrong reading, and because human input was needed at every step – each time with the risk of error.

Babbage's idea was to create a calculating machine that worked completely automatically, and so do away with human error. He wasn't the first to think of the idea, but he was the first to try and make it a practical reality. Babbage's machine was called the Difference Engine because it cleverly allowed complex multiplications and divisions to be reduced to additions and subtractions, or 'differences', that could be handled by scores of interlocking cogs.

Although work on the Difference Engine was halted after ten years, when the government withdrew funding, Babbage was undeterred and threw himself into the development of a far more sophisticated machine, named the Analytical Engine.

The Difference Engine was essentially just a clever mechanical calculator, although it incorporated such sophisticated ideas as automatic printing of results. But experts who have studied Babbage's papers believe the Analytical Engine could have truly been what we now call a computer – a machine that could 'think', responding to new problems and devising its own way of solving them without human intervention. In working out his ideas for the Analytical Engine, Babbage anticipated virtually all the key design elements of the modern computer, including the central processing unit and different kinds of memory.

Remarkably, Babbage's ideas were not just vague concepts, but thoroughly practical ideas that were simply beyond the technology of the day to build. Later in life, Babbage designed a simpler version of the Difference Engine. His design drawings for Difference Engine No. 2 were so thorough that in 1991, after 150 years, London's Science Museum were able to use them to build a full-scale version – and show that it really worked. The chances are that the Analytical Engine would have worked, too, had it been built, and the era of Charles Dickens would have had computers.

Young Babbage

Born in south London on Boxing Day 1791, Babbage was the child of wealthy parents who could afford to have him educated at some of the best private schools. His relationship with his father was never a happy one – Babbage wrote that he had a 'temper the most horrible which can be conceived' – and Babbage was left with an insecure defensive streak which was to plague him all his life.

By the age of 19, when he went to Cambridge University, he was starting to prove himself a brilliant mathematician. He upset the university authorities with a provocative final presentation, 'God Is a Material Agent', which cost him the top honours and caused him to be barred for a long time thereafter from access to academic posts. Putting this setback blithely behind him, Babbage married young Georgina Whitmore against his father's wishes and set up home in London.

Fortunately, Babbage was likeable, talented and energetic, and soon became a leading light in London scientific circles, helping to found both the Royal Astronomical Society, and the Analytical Society to promote analytical calculus and statistics. Thus, when he applied for government funding to build a full-size version of the Difference Engine after a small-scale trial in 1821, he found he had plenty of influential friends and supporters willing to vouch for his credentials.

Vive la différence

Difference Engine No. 1 was a tremendously ambitious project. No calculator had ever worked with numbers bigger than four digits, yet Babbage planned to build a machine that could handle numbers of up to fifty. Once set, it would work through the entire calculation automatically. To build it, Babbage hired Joseph Clement, perhaps the best machinist in London, set up a special dust-proof workshop and went on extensive research trips, often incognito, around the mills and workshops of the north to become familiar with the most advanced manufacturing techniques.

Each number in the Difference Engine was represented by a column of cogwheels, and each cogwheel was marked with digits from 0 to 9. A number was set by turning the cogwheels in the column to show the right digit on each. The working model had seven number columns, each of sixteen digit cogwheels or digits. Altogether, the Difference Engine had 25,000 moving parts, and many of them had to be totally identical or the machine simply would not work.

The Analytical Engine

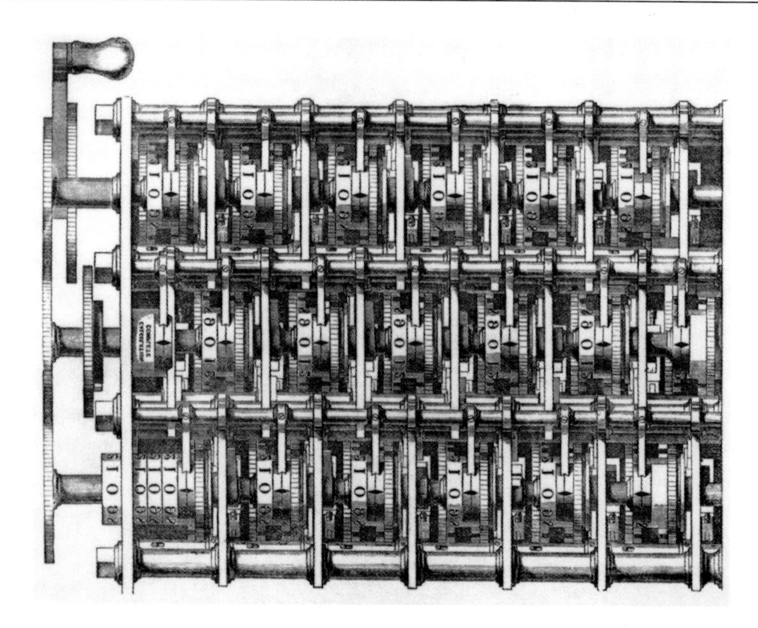

An illustration of the portion of Babbage's Difference Engine which was actually built.

Babbage's Analytical Engine anticipated many of the key design features of the modern computer by well over a century. One of the problems with a calculating machine, for instance, was what to do with carry overs. When adding up, you add each column of digits, then carry over the appropriate amount. At first, Babbage achieved this task by including a separate mechanism for each carry over. But the mechanism began to get very complicated – until he had the brilliant idea of separating out the addition and carry over processes. He split his machine into two parts: the Mill, where various arithmetical operations were performed, and the Store, where numbers were held, and to which results from the Mill were returned after processing. There could be no clearer analogy with a modern computer's central processing unit (the Mill) and its memory (the Store).

Another idea was what today we would call a program. Babbage was inspired by the Jacquard loom, which used cards punched with holes to tell a weaving machine how to weave complex patterns in silk. In 1836, he realized he could use punched cards not only to control his machine, but to record results and calculation sequences permanently. Babbage's punch cards were both the programs and the portable data storage devices of modern computers.

Science historians who have studied Babbage's papers on the Analytical Engine in recent years have been amazed by just how many features and problems of the modern computer Babbage anticipated. Yet in some ways, his impact on the development of computers was limited. All his work lay hidden in archives until well after the computer was developed – and the public percepton that the Difference Engine was a giant white elephant may have put off governments from investing in calculating machines for more than a century.

No one had ever attempted to make a machine that needed so many parts or such precision, and it pushed the machinists' skills to the limits. Like space programmes today, the project had many technological spin-offs, not least of which was the Whitworth screw, developed 20 years later by Joseph Whitworth, one of the workers on the project. The Whitworth screw was the first standardized screw system, and it revolutionized engineering.

Slow progress

The need to develop technology as he went along meant that progress was slow, and after 10 years of intense effort only half the parts had been made. In 1832, to show progress, Babbage had a small section of the Engine assembled, which he would use to entertain people, and convince them of his theory that a rational universe could spring surprises by operating according to laws which remained invisible until they suddenly took effect – a crucial notion in the theory of evolution.

By this time, Babbage had not only put a great deal of work into the project, but also £6,000 of his own money. The government meanwhile had spent £17,000 – the price of 25 of the most advanced steam locomotives of the day – and were beginning to get seriously frustrated by the apparent lack of progress. When a dispute arose between Babbage and Clement over bills, the government decided to call a halt to the project.

Hard times

The 5 years prior to 1832 had been particularly difficult for Babbage. In 1827, his father had died, followed by his son and then his beloved wife Georgina. The publication of his vocal attack on the cliquey, moribund nature of the scientific establishment had earned him a host of powerful enemies. And accusations that he was using the project to line his pocket hurt deeply. As the project ground to halt, Babbage, for the first time in a decade, had time to reflect, and as he did so, he began to realize he could go much further with mechanical calculation. He started work on a grand new idea: the Analytical Engine.

Having finished the design of the Analytical Engine on paper, and with the government still refusing to resume the funding for the Difference Engine, Babbage was at a standstill. He therefore turned his constantly inventive mind to other ideas, and every now and then accepted invitations to act as consultant on other projects. The railway engineer Brunel, for instance, asked him to assess whether broad gauge or narrow gauge was better. While working on the railway, Babbage invented a device for uncoupling carriages automatically in case of derailment, a cowcatcher, and a kind of 'black box' recorder for trains.

The 'Enchantress of Number'

Work on the Analytical Engine, though, was lonely, and Babbage's ideas were little understood, let alone appreciated. So it was with gratitude that in 1843 he gained a fan in the shape of 27-year-old aristocrat Ada Lovelace, daughter of the poet Lord Byron. Ada was convinced of her mathematical prowess, writing to Babbage that 'the more I study, the more insatiable I feel my genius for it to be'. Babbage was flattered and called her the 'Enchantress of Number'. Ada decided to take over the publicizing of Babbage's ideas and arranged for the publication of an Italian pamphlet about the Analytical Engine, to which she added extensive explanatory notes.

One key idea that emerges in Ada's notes is the notion that the Engine might have far wider applications than purely mathematical ones. It is clear that she and Babbage foresaw the real scope of mechanical brains. Ada wrote, 'Many persons ... imagine that because the Engine is to give its results in numerical notation, the nature of its processes must consequently be arithmetical and numerical.... This is an error. The engine can arrange and combine its numerical quantities exactly as they were letters or any other general symbols.'

Ending all differences

Ada's plans for Babbage were tragically thwarted when she contracted cancer and died at the age of just 36. It seems likely, though, that many claims about her contribution to the Analytical Engine are exaggerated, since Babbage's conception was virtually complete by the time she made contact with him.

Ironically, Ada's publicizing of the ideas for the Analytical Engine may have helped finally scupper the chances of further funds for the Difference Engine. It seemed to the government that Babbage himself had rendered the Difference Engine obsolete with his work on the Analytical Engine, as Prime Minister Robert Peel made clear in a stormy interview with Babbage at the close of which Babbage turned on his heel, saying indignantly, 'If those are your views, then I wish you good morning!' Afterwards, Peel commented acidly that perhaps Babbage's machine 'should be used to calculate the time at which it would be of any use'.

Taking advantage of all his work on the Analytical Engine, Babbage came up with a much simpler, more elegant design for the Difference Engine. This design, called Difference Engine No. 2, was the machine built from his plans by the Science Museum in 1991.

A reconstruction of Charles Babbage's Difference Engine. This machine was built at London's Science Museum in 1991, and works perfectly.

Years of isolation

By the 1850s, the years of disappointment were beginning to tell. Babbage had made enemies of many powerful people, and had conducted a bitter public feud with Sir George Airy, the Astronomer Royal, and Airy's friend, the astronomer Reverend Richard Sheepshanks. It was Airy's condemnation of Babbage's work that had finally prompted the government to stop funding. Babbage accused him of envy and malice; his feud with Sheepshanks was even nastier.

The Swedish Babbage

Babbage – now in his sixties – was beginning to lose heart. Then, in 1854, a Swedish printer called Georg Scheutz brought to London a rough-and-ready difference engine he had developed himself, inspired by reading about Babbage's 20 years earlier.

Scheutz had been worried how Babbage might react to a rival, but Babbage welcomed him with open arms, helping him find contacts in London to promote it. It was Scheutz who predicted in 1857: '[Babbage] will then be known for what he truly is – namely one of the benefactors of mankind, and one among the noblest and most ingenious of the sons of England.'

Yet Scheutz's machine, like Babbage's, aroused no more than curiosity, since printed tables and human computers remained a cheaper, more practical option. Babbage resumed work on the Analytical Engine, and wrote insightful papers on a huge range of scientific topics, including ciphers and codes, chess, lock-picking, industrial economics, geology, submarines, and astronomy.

Babbage was ageing, and his nights were beset by nightmares, hallucinations and dreadful headaches. He was also plagued by street musicians outside his house, suffering an ear condition which made their playing extremely upsetting to him. In October 1871, he fell seriously ill, and on the 18th, with the organ players in the street playing as loud as ever, he died.

Charles Darwin

Few ideas in the history of science have so completely altered the way we see ourselves as Charles Darwin's (1809–82) theory that all life, including humans, evolved into its present form through a process of natural selection.

1809–82

In the century before Darwin was born, scientific observation and the powerful rationality of the Age of Enlightenment were slowly changing the way people looked at the natural world. It was no longer considered quite so mysterious and magical, but something to be catalogued, studied and probed with help from the growing body of scientific knowledge.

Throughout the latter part of the eighteenth century, botanists had built on Linnaeus's work, discovering and classifying more and more species of plants around the world. To a lesser extent, zoologists had done the same thing with animals. Great new botanical gardens, such as Kew in London, and zoological gardens were the living testament to the efforts of these species hunters.

Both scientists and theologians began to ask just how all these species had come about, and why each seemed so perfectly suited to the environment in which it lived – fish for swimming in the sea, birds for flying in the sky, and so on.

The orthodox view was that of the Creationists. According to the book of Genesis in the Bible, 'God created every living creature that moves ... every winged fowl ... and every thing that creeps on the face of the Earth'. So the Creationists believed, as many still do, that every species was created at once by God – and that each was perfectly designed by him suit the conditions in which it lived. In 1802, the theologian William Paley argued in favour of this original design idea with an example: if you found a watch in the desert, you would surely assume it had been made by some skilled watchmaker. How much more skilled, then, was the watchmaker who fashioned the human eye?

However, some thinkers were beginning to question the idea that all species have been there from the start, unchanging. More and more naturalists were looking at fossils and finding they were of species that often seemed very different from those alive today. Where had those species gone, and why were there so few fossils of creatures that are alive today?

At the same time, geologists such as James Hutton were beginning to challenge the orthodox idea that the world was just a few thousand years old, and that all the landscapes were created in a series of brief catastrophes. A growing minority were arguing that the Earth is in fact very old, and that landscapes were created by long slow cycles of erosion and upheaval.

Against this background, more and more thinkers began to argue that species are not fixed but have actually changed, or evolved, through time. One of these thinkers was Charles Darwin's grandfather Erasmus Darwin. Another, and perhaps the most famous, was the French naturalist Jean-Baptiste Lamarck.

Lamarck not only developed a picture of how species progressed in a purposeful manner from a single-celled organism to the supreme species, mankind, but also suggested just how this evolution takes place. He argued that each species has an 'inner feeling' which propels it to ascend the ladder of evolution. He also argued that skills which aid survival can be passed on to the next generation and so gradually built up. A giraffe that stretched its neck to reach higher branches, for instance, would pass on its long neck to its offspring.

Lamarck's ideas so shocked those with orthodox religious views that those championing them were often vilified and sacked from teaching positions – even as late as 1840. Many scientists, too, found the idea of inheriting acquired characteristics unconvincing. And Lamarck's system hardly began to explain how each species is so marvellously adapted to its environment.

Darwin's great breakthrough was not the discovery of evolution – Lamarck and others had done that. What he did was work out what exactly evolution is and how it happens. His insight was to focus on individuals, not species, and he showed how individuals evolve by natural selection. Natural variation within a group of individuals means that some will be better equipped to survive in particular conditions, and if they survive they will pass their characteristics on to their offspring. Later commentators have characterized his idea as 'survival of the fittest', but it was never a phrase Darwin himself used. This mechanism explained how all species, including humans, evolved to become well suited to their environment.

Young Darwin

Darwin was born on 12 February 1809 in Shrewsbury, the son of a well-to-do country doctor. Charles was the youngest of the family and the only boy, and throughout his childhood he was doted on by his sisters. He was educated at the local public school, but was too busy collecting nature specimens and conducting chemistry experiments to shine in class. 'The school,' Darwin said, 'as a means of education to me was simply a blank.'

At the age of 16 he went off to Edinburgh to study medicine like his father. But Darwin found all the operations far too gruesome and spent much of his time with the zoologist Robert Grant, a great believer in Lamarck's ideas. Both were avid collectors and spent many a day rambling in the Scottish hills looking for plants.

Since he was clearly not suited for the study of medicine, Darwin's father sent him off to Cambridge to study divinity at Christ's College. Here again, Darwin was distracted by another naturalist – this time the Reverend Professor John Henslow, who had energetically restored Cambridge's botanical gardens after years of neglect. As with Grant, the two of them formed a firm bond and would often go specimen hunting together.

The trip of a lifetime

In 1830, Henslow was offered the post of botanist on HMS *Beagle*, soon due to set sail to South America on a surveying trip for the Admiralty. Henslow was unable to

Charles Darwin

go, but offered the job to young Darwin instead. At first his father refused permission, but he relented after strenuous efforts from his daughters.

The *Beagle* voyage was to be, quite literally, the trip of a lifetime for Darwin (see box on p. 94). Initially meant to take two years, it actually lasted over five. By the time Darwin came back, he was a changed man. Not only had he gathered enough data about species around the world to last him a lifetime, but he had learned enough by close observation of the extraordinary range of wildlife he saw to start sowing the seeds of his theory of evolution. Nonetheless, Darwin was never a hasty man, and it took him many years of patient thought and study before his theory was ready for publication.

Fame and marriage

In the meantime, he found himself quite a celebrity when he returned to London, for Henslow had been giving lectures based on the specimens and notes Darwin had sent back whilst sailing round the world. He was appointed a fellow of the Geological Society, invited to join the exclusive gentleman's club, the Athenaeum, and elected to a fellowship of the Royal Society. Yet he was never one to seek the limelight, and over the next few years he spent much of his time quietly making notes to develop his ideas on the species question, building up data, visiting zoos, talking to plant breeders, naturalists, birders and anyone who might give him some more background information.

Although he loved the quiet, studious life, he also began to feel the need for companionship, and in 1838 the 29-year-old Darwin decided to marry his cousin Emma, having carefully weighed the pros and cons. It proved to be a very happy marriage, and the couple soon moved to Down House near Bromley in Kent and remained there for the rest of their lives. Darwin was never in robust health – and may have caught some tropical disease on the Beagle voyage – but he was well looked after at Down House by Emma, and continued building up his ideas on evolution.

The breakthrough

Although Darwin's ideas mostly grew by slow accumulation, there was one 'Eureka!' moment, when he read *An Essay on the Principle of Population* by Thomas Malthus. In this essay, written in 1798, Malthus had argued that populations – both human and animal – will always multiply until they exceed the amount of food available, at which point the population will crash, only for the process to start again. Darwin was excited: 'It at once struck me that under these circumstances favourable variations would tend to be preserved, and unfavourable ones to be

The title page of Darwin's Origin of Species. *Published in 1859, the first print run was sold on the first day; the controversy it generated, however, would last far longer.*

ON

THE ORIGIN OF SPECIES

BY MEANS OF NATURAL SELECTION,

OR THE

PRESERVATION OF FAVOURED RACES IN THE STRUGGLE FOR LIFE.

BY CHARLES DARWIN, M.A.,

FELLOW OF THE ROYAL, GEOLOGICAL, LINNÆAN, ETC., SOCIETIES;
AUTHOR OF 'JOURNAL OF RESEARCHES DURING H. M. S. BEAGLE'S VOYAGE ROUND THE WORLD.'

LONDON:
JOHN MURRAY, ALBEMARLE STREET.

Darwin's desk at Down House, the country retreat where he moved in 1838, and where he would spend the rest of his life, insulated from the storms his work would arouse.

destroyed. The result would be the formation of new species.... I had at last got a theory by which to work.'

Yet though he had his theory, he kept it very much to himself. Indeed, he soon embarked on ten years' work, writing a treatise on a single species of barnacle, which he described thus: 'Mr Arthrobalanus [is] an enormous coiled penis.' He might have carried on in this way if a bombshell had not dropped suddenly on his desk in the summer of 1858.

The rival

The bombshell was a letter from Alfred Wallace, a young naturalist then lying ill with malaria in the Molucca Islands in Indonesia. Wallace outlined a theory of evolution by natural selection which corresponded almost exactly to Darwin's own. 'I never saw a more remarkable coincidence,' Darwin later commented. Darwin talked to his friends – the famous geologist Charles Lyell, the botanist Joseph Hooker and the philosopher T.H. Huxley – and together they arranged for Darwin's and Wallace's ideas to be presented together, making it clear that Darwin had developed his ideas 12 years previously.

Spurred into action, Darwin then wrote his great book *On the Origin of Species*, in which he outlined his ideas and gave a wealth of supporting evidence gathered from the *Beagle* voyage and subsequent research. It was a sensation, and the first edition of 1,250 copies sold out on the day of publication, 24 November 1859.

The great debate

Some people immediately embraced the idea, seeing how it explained a huge amount about the natural world. Others condemned it as an affront to God, because nowhere did Darwin's ideas leave room for the biblical Creation. Heated discussions began to take place around dining tables and in debating chambers across England.

The most famous encounter was between Darwin's friend T.H. Huxley and the Bishop of Oxford, 'Soapy' Sam Wilberforce. At one point in the debate, Wilberforce challenged Huxley to say whether it was on his grandfather's side or his grandmother's that he was descended from an ape. But this cheap jibe cost Wilberforce victory. Turning it neatly round, Huxley argued persuasively and seriously enough to carry the day. This was the picture across the country, and the Darwinists, as they came to be called, gradually won more and more people to their cause.

A major setback occurred in 1862, when the Scottish physicist William Thomson, later Lord Kelvin, estimated the age of the Earth scientifically. Kelvin declared the Earth could be no older than 40 million years, and possibly only 20 million years. His calculation was based on how long it would have taken the Earth's interior to cool down to its current temperature from its original molten state. This was a real blow, because Darwin's theory depended on the Earth being much older. But it turned out that Kelvin was mistaken about how fast the Earth is cooling; further calculations showed that the world is over 4 billion years old.

Human descent

In the meantime, Darwin, who kept quietly out of the debate down in Down House, wrote *The Descent of Man* (1871), in which he explained how his theory of evolution applied to the evolution of mankind from the apes. In a famous passage, Darwin wrote, 'Man with all his noble qualities ... still bears in his bodily frame the indelible stamp of his lowly origin.'

Darwin went on developing his ideas, particularly in relation to humans, for the rest of his life. In 1872, at the age of 63, he published an important book on how emotions and expressions might have evolved, entitled *Origin, the Expression of the Emotions in Man and the Animals.*

By now, the long years of poor health and hard study were taking their toll. He died on 19 April 1882 aged 73 and was widely mourned. He was buried with honour in Wesminster Abbey, his coffin carried by, amongst others, his friend and champion T.H. Huxley.

A Galapagos finch, one of the birds which helped inspire Darwin's theory of evolution.

The voyage of the *Beagle*

In his autobiography, Darwin wrote, 'The voyage of the *Beagle* has been by far the most important event in my life and has determined my whole career.' When Darwin embarked on the voyage, he was simply an amateur botanist with only a basic knowledge of plants, very little zoology and no geology. As Darwin set sail, Henslow gave him a copy of the first volume of Charles Lyell's newly published book *Principles of Geology*. It was a revelation to him, showing how landscapes had evolved gradually through long cycles of erosion and upheaval. At the *Beagle*'s first stop in the Cape Verde islands, Darwin saw a seam of white coral running up the side of the volcano, showing that it had gradually been uplifted from the sea, not suddenly as the Catastrophists said. It was enough to convince him that Lyell's ideas were right, and he wrote home, 'Geology carries the day.'

As the voyage continued down the east coast of South America, Darwin gathered a huge number of specimens and took copious notes on the wildlife and geological features that he encountered. With everything he saw, the idea that species are designed and fixed once and for all seemed increasingly unlikely to him. The culmination of his observations came in the Galapagos Islands off the west coast of South America in autumn 1835. There are twenty or so islands here and each one, Darwin noted, had its own subspecies of finch with its beak perfectly adapted to its way of feeding. Some used their beaks to crack nuts, others to suck nectar from flowers, and so on. As Darwin said, 'One might really fancy that from an original paucity of birds in this archipelago, one species had been taken and modified for different ends.' It was for him the clinching piece of evidence that species evolved to suit their environment; they were not 'designed' to suit it right from the start.

By the time Darwin came home, he was firmly convinced of the reality of evolution. All he needed to do was to work out how and why it happened. That was his lifetime's work.

Louis Pasteur

With his brilliant work on microbes, the great French scientist Louis Pasteur (1822–95) pioneered the theory that germs cause disease, discovered how to vaccinate against rabies, anthrax and chicken, and found the method of treating wine, beer and milk to prevent them spoiling that now bears his name.

1822–95

In the mid-nineteenth century, although trains were speeding people quickly from place to place; telegraphs were enabling instant communication; and scientists were discovering fields, radiation and subatomic particles, medicine was still comparatively in the dark ages. The chances of surviving childhood, even among the well-to-do, were not that good. Victorian families almost expected at least one of their children to die young, though that did not make it any easier to bear. Women knew there was quite a high chance they would die giving birth, especially if they went into hospital. After a surgical operation, the chances were that disease would kill the patient, even if the surgery did not. And diseases like cholera and consumption (TB) took a high toll, especially among the poor.

Doctors were virtually powerless against many illnesses, and catching diseases like consumption or syphilis was almost certainly a death sentence. Doctors had very few of the drugs we take for granted today. Indeed, the only effective drug they had was opium, to kill pain. They also had no idea what actually caused infectious disease.

Leeuwenhoek had hinted at the possible role of microbes in disease when he discovered them in the seventeenth century, but no one had really taken up the idea. In the nineteenth century, many doctors still believed that diseases were caused by miasma, a mysterious toxic vapour that welled up from stagnant waters, slums and faeces. Even though they had observed bacteria in wounds and diseased tissues, they believed they appeared there spontaneously – that they were actually a natural result of any decaying process.

Pasteur's initial interest in microbes came from his research into wine-making, which showed that micro-organisms called yeasts are what make wine ferment and mature. He discovered how the wrong kind of microbes made wine go off – and that they could be eradicated by heat treatment. He then went on to prove that spontaneous generation – the idea that microbes appear from nothing – was a fallacy. Pasteur became convinced that germ microbes cause disease, which was an idea also suggested by the young German scientist Robert Koch, and Pasteur went on to show how many diseases could be prevented by arming the body's defences through vaccination with weakened forms of these germs. Over the next few decades, scientists succeeded in identifying the germs responsible for a range of diseases including TB, cholera, diphtheria, leprosy, tetanus, malaria and yellow fever.

The young artist

Pasteur was born on 27 December 1822 in Dôle, Bourgogne. His father, Jean-Joseph, was a tanner, and only reasonably well off. As a boy, Louis had more talent for art than science, and it was said that he could have been a great painter. However, as he grew older, he showed an increasing interest in science, and when he was 21 he entered the prestigious École Normale Superieur to train as science teacher.

A year after graduating, he presented his first paper to the Academy of Sciences. It was a brilliant debut. In it he explained how it is that crystals of tartaric acid – which are made when grapes ferment – and crystals of racemic acid are chemically identical but polarize light differently and have different properties. Pasteur showed that their crystals were not actually identical, but perfect mirror images of each other.

Such was the impact of Pasteur's paper that he was given the *Légion d'honneur* (a French order of merit) by the French government and a gold medal by the English Royal Society. Pasteur later came to be recognized as the founder of what is now known as stereochemistry.

The workaholic professor

He was made Professor of Chemistry at the University of Strasbourg because of these honours, and married Marie Laurent. As he did throughout his life, he began to work extremely hard at his research. Later in life, Marie wrote to her daughter, 'Your father, very busy as always, says little to me, sleeps little, and gets up at dawn – in a word, continues the life that I began with him 35 years ago today.'

In 1854, Pasteur was made head of science at the University of Lille. Because he believed teachers and industrialists should not live in separate worlds, he began to make quite a name for himself in the city by organizing classes for working men and trips to factories for his science students. Encouraged by this, the manager of a local vinegar factory approached him with a problem. The factory made vinegar from beet juice, but the fermentation process often seemed to go wrong, spoiling the vinegar. Similar problems occurred in wine- and beer-making.

Discovering yeast

Previously, scientists had assumed fermentation was a chemical process, but when Pasteur investigated with the aid of a microscope, he saw that when wine and beer age properly, they contained tiny round microbes called yeast. It was these microbes, Pasteur realized, that made the alcohol in fermentation. He also observed that wine and beer contained long and narrow yeast cells when the process went wrong. It was clear to him that there were two kinds of yeast involved – one that made alcohol and another that made unwanted lactic acid.

In the way that came to typify his approach, Pasteur was not content with discovering the problem; he wanted to find a solution. He soon found that he could kill off the harmful yeast while leaving the good yeast undamaged by heating the wine gently to about 140° F (60° C). The wine-makers were at first sceptical of his

findings, and were worried about the effect that heating would have on flavour, but they were soon won over, and the process of 'pasteurization' is now widely used to make wine, beer, milk and many fruit juices safe to drink.

Copper distilling equipment, originally belonging to Louis Pasteur, and now on display in the Pasteur Museum in Paris.

Microbes in the air

Pasteur's work with yeast cells made him begin to wonder how such microbes appeared in the first place. He was convinced they were not spontaneously generated as many people said. By a series of simple but ingenious experiments, he showed that food goes off when in contact with ordinary air, but not when exposed only to air which has been heated to kill off any microbes. He also demonstrated that food did not go off so quickly high up in the Alps where the air is thin. This proved that the mould on bad food is not generated spontaneously by the food but comes from spores in the air – the fewer spores there are in the air, the more slowly food will go off.

Pasteur was beginning to acquire a Mr Fix-it reputation, and in 1865 he was asked to investigate the disease that was killing silkworms and threatening the silk industry of southern France. Reluctant at first because he knew nothing of silkworm caterpillars, he finally agreed and studied the problem with his usual vigour. He quickly realized that a tiny parasite was the culprit and recommended a drastic solution – destroying all the infested worms and mulberry trees and starting again. The silk-makers took his advice, and the silk industry survived.

His reputation was now so great that in 1867 Emperor Napoleon III had a special laboratory created for him to replace his attic laboratories at the École Normale, which he had to enter on his knees. Unfortunately, he only enjoyed his new space for a year when a stroke left him permanently paralyzed in his upper arm and left leg. He was thereafter dependent on laboratory assistants to perform all but the simplest laboratory work for him.

Microbes and disease

By now, Pasteur was becoming convinced of the role of microbes in infection. The English surgeon Joseph Lister had read about Pasteur's work and had realized that surgical operations could be made much safer if wounds were cleaned and dressings sterilized to destroy microbes. The death rates in operations began to drop at once after this 'antiseptic' procedure was introduced. Lister recognized Pasteur's contribution, saying at Pasteur's seventieth birthday celebrations, 'Truly there does not exist in the world an individual to whom medical science owes more than you.'

Pasteur himself began to investigate the way diseases are spread among humans and animals, and was sure germs played a part. In 1876, Robert Koch claimed to have detected the germs that caused the sheep disease anthrax. Pasteur ran his own tests and not only confirmed Koch's findings but showed that the germs can survive long periods in the soil. Therefore healthy sheep could easily pick up the disease from a field previously occupied by sick sheep.

The power of inoculation

At first, Pasteur thought to treat the problem as he had the silkworm infestation – by killing infected sheep and burning the land. Then he realized that any sheep that survived a bout of anthrax became immune to the disease. A century earlier, Edward Jenner had shown how inoculating people with cowpox, a mild version of the smallpox disease, protected them against smallpox. Pasteur wondered if this might work for anthrax. In a famous experiment, he showed that sheep inoculated with anthrax germs weakened by heating survived inoculation with full-strength germs, while untreated sheep did not.

In this way, millions of sheep were protected against anthrax, and in 1878 Pasteur showed that inoculation worked for chicken cholera, too. Soon Pasteur turned his attention to rabies, and his success in creating a vaccine against this terrible disease made him famous across Europe. Indeed, it caught the public imagination, and a movement to collect funds to further his work had contributions flooding in from far and wide – from both ordinary people and royalty, including the Czar of Russia. Over 2 million francs were raised, and on 14 November 1888 the Pasteur Institute opened in Paris.

Pasteur died on 28 September 1895 with the words: 'One must work; one must work. I have done what I could.' He was given a full state funeral and buried in a magnificent tomb in the Pasteur Institute, its walls decorated with paintings of the lambs, dogs and children whose lives he had saved.

Preventing rabies

In the 1880s rabies was a particularly nasty major disease that killed its victims slowly and agonizingly, and Pasteur was determined to find a way to beat it. At great risk to his life, he took samples from the jaws of rabid dogs by sucking the saliva through a glass tube and infected rabbits with them. In this way, he created a weakened version of the virus by drying out the rabbits' spinal cords after they died. When this weakened virus was injected into healthy rabbits, it protected them from rabies.

Pasteur was unsure whether this vaccine would work on humans, and was wary of trying. Then, in July 1885, a 9-year-old shepherd boy called Joseph Meister was brought to him after being bitten fourteen times by a rabid dog. If Pasteur did nothing, the boy was certain to die in agony, so he took the chance and injected the boy with his rabies vaccine. Fortunately the treatment worked, and the boy remained healthy.

As news of Pasteur's success spread through Europe, he found people turning to him for a cure. Before long, a party of nineteen Russians who had been bitten by a mad wolf turned up on his doorstep. It was two weeks since they had been bitten, and Pasteur feared the disease might have gone too far. Again, though, his vaccine worked, and sixteen of them survived. Over the next 10 years 20,000 rabies victims were given Pasteur's treatment, and only 200 died.

Gregor Mendel

A provincial nineteenth-century Augustinian monk, Gregor Mendel (1822–84) was perhaps an unlikely pioneer in the field of life sciences. Yet he is now regarded as having laid the foundations of the modern science of genetics.

1822–84

GREGOR MENDEL WAS THE FIRST PERSON TO USE MATHEMATICS and applied statistics in biology, and his principles of heredity would revolutionize the cultivation of plants and the breeding of livestock in the twentieth century.

Mendel was born in July 1822 in Hyncice, a remote area of the Austrian Empire in what is now the Czech Republic. The young Mendel performed well at school, showing a strong interest in natural science. In 1843 Mendel entered an Augustinian monastery at Brünn, Moravia (modern-day Brno, Czech Republic) and began training for the priesthood. He was ordained in 1847.

In the course of his training, he also found time to teach himself some science, and in 1849 he worked briefly as a substitute teacher of mathematics at a nearby secondary school. The following year he failed his teaching exam, ironically receiving his lowest marks in his biology paper. In 1851 the abbot sent Mendel to the University of Vienna to study physics, chemistry, mathematics, zoology and botany. Three years later he returned to Brünn and taught natural science at the local high school. He continued teaching until 1868, when he was elected abbot of the monastery. He never did pass his teaching exam.

A fascination for plants

The research that would establish Mendel's place in scientific history began in the small monastery garden at Brünn in 1856. Mendel had always loved nature, an interest first aroused by his experience on his father's orchard and farm. He was fascinated by plants and often wondered how they obtained atypical characteristics. The biologist Lamarck (1744–1829) had suggested that plants were influenced by their environment, and Mendel wanted to test out this theory. During one of his regular walks around the monastery garden, he found an atypical variety of an ornamental plant. He replanted it next to a typical variety, and then studied their offspring. The offspring retained the essential traits of their parents, which meant that they were not influenced by their environment. This simple test set Mendel on the path that would lead to his discovery of the laws of heredity.

Mendel was helped in his research by the scholarly atmosphere that prevailed at the monastery and at the high school where he worked. The monastery's abbot was an enthusiastic amateur botanist, and several of Mendel's work colleagues were also interested in science. In 1862 a few of them founded the Natural Science Society, and Mendel played an active role in the society's meetings. Furthermore, the libraries of both monastery and school contained many works of science, especially on his favourite subjects of agriculture, horticulture and botany.

Garden peas

Emboldened by his initial research, Mendel embarked on a long and rigorous series of experiments on garden peas, using a greenhouse in the monastery garden. First, he spent two years preparing his specimens – seven varieties of pea – to make sure they bred true. Each one was bred for a particular characteristic, such as tallness or shortness, differences in seed colour or pod shape, and the position of the flowers on the stem. Then, helped by two assistants, he repeatedly cross-bred these varieties – 30,000 pea plants in all. Mendel's experiments on peas took a total of seven years. They took so long mainly because of his determination to be as accurate as possible in his work. He took great pains to avoid accidental cross-fertilization and meticulously noted down every tiny variation in the plants' offspring.

Mendel was not the first to experiment with plants in order to understand the nature of inheritance. However, his experiments differed from previous research in two important respects. Firstly, instead of looking at the characteristics of the whole plant, Mendel focused on single, clearly visible and distinguishable traits, such as round versus wrinkled seeds or purple versus white flowers. Secondly, he made precise counts of the number of plants bearing each trait. This quantitative data allowed him to see statistical patterns and ratios that had eluded his predecessors.

Mendel observed that the first generation of hybrids (crossbred plants) usually showed the traits of only one parent. For example, the crossing of yellow-seeded plants with green-seeded ones gave rise to yellow seeds, and the crossing of tall-stemmed plants with short-stemmed ones gave rise to tall-stemmed plants. Mendel was therefore able to conclude that certain traits, such as yellow seeds and tallness of stem, were dominant, and other traits, such as green seeds and shortness of stem, were recessive. At first it appeared that the dominant traits consumed or destroyed the recessive traits. But Mendel knew that this could not be the case when he observed that the second generation of hybrids exhibited both the dominant and recessive traits of their 'grandparents'. Furthermore – and this is where Mendel's accurate counting really helped – these traits reappeared in consistent proportions in each experiment. About three-quarters of the second generation plants showed the dominant trait, and a quarter showed the recessive trait.

From this Mendel concluded that each parent plant carries a pair of determining 'factors' for each trait. In other words, it carries a pair of traits for height of stem (tall and short), seed colour (yellow and green), and so on. He realized that these pairs of factors are passed onto their offspring during reproduction, and that one trait in the pair can sometimes dominate the other. What Mendel called 'factors' are now known as genes, although the term was not coined until 1909.

Mendel deduced that these factors do not blend or mix with each other – the offspring of yellow-seeded and green-seeded parents don't have yellowy green seeds – but remain pure and uncontaminated. And when the hybrid plant forms its reproductive cells (or gametes), the genes segregate and pass to different gametes. Thus, an offspring inherits from a parent either one trait or the other, but not both.

This is known as Mendel's first law, or the principle of segregation. By applying this law across several generations, Mendel was able to predict accurately the number of offspring exhibiting each trait.

He also tried cross-breeding pea plants that differed in two or more traits. He found that their traits reappeared in every possible combination in their offspring – wrinkled seeds with fat pods, smooth seeds with thin pods, and so on. The segregation of pod shape occurred independently of the segregation of seed surface, and the traits combined with each other at random. This is known as Mendel's second law, or the principle of independent assortment.

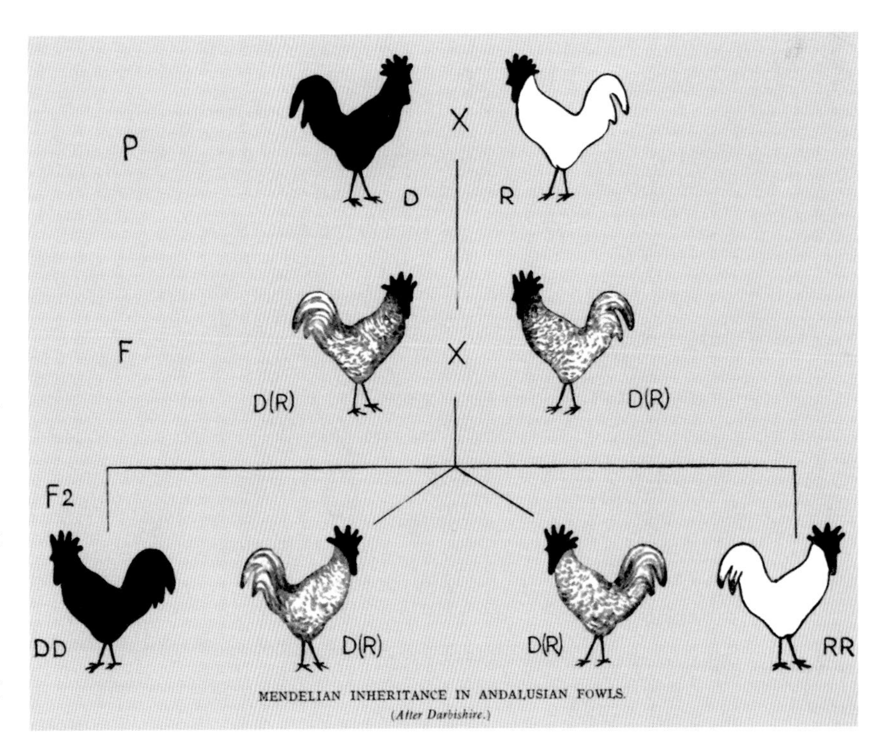

Mendel's principles of heredity are displayed here using the example of chickens. P represents the parents; F1 and F2 are the first and second generations. D is the dominant gene, and R the recessive.

Publishing his findings

He confirmed these biological principles by conducting similar experiments on flowers, corn and other plants. Then, eight years after he had first embarked on his research, Mendel decided he was ready to share his findings with the scientific community. He presented the results of his experiments in two papers at meetings of the Natural Science Society in Brünn in February and March of 1865. He received a polite hearing from the forty or so in the audience, yet no one present appeared to acknowledge that his discoveries broke new ground. His methods were simply too unusual to be appreciated. No one before him had attempted to use mathematics and statistics as a means of understanding and predicting biological processes. Mendel was also a shy character, more at home in the potting shed than the lecture hall, and may not have delivered his paper with the necessary conviction.

The Society published Mendel's article, 'Experiments with Plant Hybrids', in 1866, and it was sent out to all the major libraries in Europe and America. Despite reaching a wider audience, Mendel's work still had little or no impact. In an effort to gain greater recognition for his theory, Mendel asked a fellow monk to send out reprints of his paper to forty eminent botanists and scientists with an interest in plant hybrids. Of the forty, only one showed any real interest. He was the Swiss botanist Karl Wilhelm von Nägeli, a leading authority in the field, who was at that time teaching at the University of Munich. It is clear from their subsequent correspondence that Nägeli either did not properly read the paper, or at least failed to appreciate its significance. He told Mendel that his work was incomplete – despite the fact that Mendel had studied over 300 cross-bred strains and a total of 30,000 plants – and urged him to continue with his experiments. Nägeli also offered him some very poor advice. He suggested that Mendel try breeding hawkweed, a plant fundamentally unsuited to the study of heredity. Hawkweed belongs to a family of

The seven traits of peas

For his experiments in plant hybridization, Mendel chose a common garden pea (*Pisum sativum*). He looked for simple traits to study – now called 'Mendelian traits'. Simple traits are those which occur in one variation or another, with no in-between. The seven traits he studied in his peas were as follows:

Trait	Dominant Trait	Recessive Trait
Type of seed surface	smooth	wrinkled
Colour of seed albumen	yellow	green
Colour of seed coat	grey	white
Form of ripe pod	inflated	constricted
Colour of unripe pod	green	yellow
Position of flowers on stem	axial	terminal
Length of stem	tall	short

Mendel described the care he took in picking a plant to experiment on. They must 'during the flowering period, be protected from the influence of all foreign pollen, or be easily capable of such protection [because] accidental impregnation by foreign pollen ... would lead to entirely erroneous conclusions'.

plants that breed asexually – the offspring is formed from an unfertilized egg. Therefore any experiments in crossbreeding hawkweed were bound to fail, since the genes of the offspring come from just one parent. Mendel did not know this – and nor did anyone else at that time – so he went ahead with a study of hawkweed, and published a paper on his unsuccessful experiments in 1869.

Mendel continued his investigations in botany and other subjects that interested him, including the study of bees, mice and sunspots, until his death in 1884. However, with his appointment as abbot in 1868, he was longer able to devote as much time to science. The task of running the monastery took up a great deal of his energies. It is also likely that he had been disheartened by the failure of his hawkweed experiments and by the general lack of recognition for his achievements. Nevertheless, perhaps Mendel did feel that his time would one day come, even if he did not live to see it. In 1883, just a few months before his death, he commented: 'My scientific studies have afforded me great gratification; and I am convinced that it will not be long before the whole world acknowledges the results of my work.'

Posthumous recognition

Mendel's prediction came true in 1900. In that year, three European botanists, Carl Erich Correns, Erich Tschermak von Syseegg and Hugo de Vries, each working independently, obtained results that showed how plant heredity was governed by a set of basic laws. Searching back through the scientific records, they discovered that a half-forgotten monk-scientist had already discovered these laws and published the results 34 years previously. At last Mendel's work was recognized.

In the early years of the twentieth century, Mendel's laws were fully tested and confirmed, and were found to have general application, not just in the world of plants, but to all organisms, including humans. Scientists continued to use the statistical methods pioneered by Mendel to explore the complex world of genetic transmission. Eventually this led them to an understanding of genetics at the molecular level, including the discovery of the DNA and RNA molecules that constitute the genetic material in all living things.

The science of genetics has transformed all our lives. Today, scientists have mapped the human genome, cloned animals, grown GM foods, and have the potential to banish inherited diseases. It is extraordinary that this great revolution in human knowledge was begun by a monk growing peas in a monastery garden, whose most valuable resources were his patience, and his care and accuracy.

Dmitri Mendeleyev

'It is the function of science,' said Dmitri Mendeleyev (1834–1907), 'to discover the existence of a general reign of order in nature and to find the causes governing this order.' And it was indeed Mendeleyev's fate to discover such a 'reign of order' with his development of the Periodic Table of Chemical Elements.

1834–1907

It is hard to overstate the importance of this achievement in the advancement of chemistry as a science. Before Mendeleyev's table, the subject was in a fair amount of chaos. New elements were being discovered, but there was little consistency in the symbols and abbreviations used, and no one could work out how to organize or arrange the elements in a way that made any sense. Mendeleyev's Periodic Table established an order and a clarity that transformed the study of chemistry. 'Without a doubt, the Periodic Table of the Chemical Elements is the most elegant organizational chart ever devised,' wrote Robert E. Krebs in his book *The History and Use of Our Earth's Chemical Elements*. Furthermore, by looking at the gaps in his table, Mendeleyev was able to predict the discovery of new elements with what proved to be extraordinary accuracy.

Early life

Dmitri Ivanovich Mendeleyev was born in 1834 in Tobolsk in the far west of Siberia, Russia. He was the youngest of 14 children (or 17 or 11, depending on which source you read). His father, Ivan, the headmaster of a local school, went blind soon after Dmitri's birth and was forced to retire. His pension was too small for the needs of such a large family, so Dmitri's mother, Maria Kornileva, had no choice but to go out to work. Her family owned a glass factory in Aremziansk, 32 km from Tobolsk, which she managed in return for a modest wage.

Dmitri, her youngest, was also Maria's favourite, and from his earliest years she began putting money aside to send him to university. Dmitri went to the local school where his father had taught, but he also received a practical education at his mother's glass factory, where he spent hours listening to the chemist and the glass blower discuss the secrets of glass making.

When Dmitri was in his early teens, the family was hit by a double calamity. In 1847, his father died, and the following year the glass factory burned down. The family was reduced to poverty. Yet the remarkable Maria would not give up her dreams for her son – she was determined that he should still have an education. In 1849, she hitchhiked with Dmitri and her other remaining dependent child, Elizabeth, all the way to Moscow – a journey of 2,000 km – in the hope of securing him a place at the university. As a Siberian, however, Dmitri was barred from entering Moscow University. Undeterred, Maria led her son and daughter a further 600 km to St Petersberg. Here again he was refused admission to the university, as well as to the medical school. Finally, in 1850, he was granted a place at the city's Pedagogic Institute. Just ten weeks later, Maria died, exhausted by her efforts. A short time later, Mendeleyev's sister expired from tuberculosis.

The young chemist

Despite these tragedies, Mendeleyev applied himself diligently to his studies, and in 1855, at the age of 21, he qualified as a teacher, winning a gold medal for being first in his class. This achievement was made even more remarkable by the fact that he had been bedridden with tuberculosis for much of his final year. Following his graduation, he obtained an advanced degree in chemistry, and in 1857 he received his first university appointment.

Around this time, the Russian government decided it was important to keep up with Western advances in science and technology, and in 1859 they paid for a number of Russian scientists, including Mendeleyev, to study in Europe. Mendeleyev spent two years at the Universities of Heidelberg and Paris. Here he met some of the leading scientists of the age, including the chemists Robert Bunsen, Henri-Victor Regnault and Stanislao Cannizzaro, and the physicist Gustav Kirchhoff. Through studying and conversing with these figures, Mendeleyev learned more about such things as the density of gases, the methods of determining the chemical composition of substances, and atomic weights – all of which greatly influenced his later work.

Back in St Petersberg, Mendeleyev continued to teach and also found time for writing. In 1861 he published *Organic Chemistry*, a prize-winning textbook that greatly raised his standing in Russian chemistry education. In 1864 he was appointed professor of chemistry at the Technical Institute, and three years later he was made professor of general chemistry at St Petersberg University – the institution that had rejected him 17 years before. Discovering that he lacked a textbook to meet his teaching needs, Mendeleyev set about writing one. The result was *The Principles of Chemistry* (1869). This became a classic work, going through eight editions in Mendeleyev's lifetime.

'Look for peace and calm in work: you will find it nowhere else. Pleasures flit by – work leaves a mark of long-lasting joy.'

Organizing the elements

It was in the course of writing this book that Mendeleyev stumbled on a discovery that would lead to his greatest achievement and would ultimately transform the whole subject of chemistry: the Periodic Table. While writing *The Principles of Chemistry*, Mendeleyev wondered if there was a logical order in which he could discuss the elements. At this time, some seventy separate chemical elements had been identified, but no system had yet been devised to order them. In this, the field of chemistry differed from other disciplines. Organic chemistry – the study of carbon-based compounds – had already been successfully organized by type, and as a result the subject could be written about and taught in a clear and systematic way. In similar fashion, biologists had found a means of classifying and ordering plants and animals. Mendeleyev wished he could do something equivalent with the chemical elements so that he could present his book with a more logical structure.

The Periodic Table

In coming up with his Periodic Table, Mendeleyev later said that he had been inspired by the card game called patience (or solitaire in North America) in which cards are arranged horizontally by suit and vertically by number. He would write the names and properties of the elements on cards and play 'chemical solitaire' with them on long train journeys, gradually filling in the gaps. Mendeleyev arranged the elements in horizontal rows called periods and vertical columns called groups. This showed one set of relationships when read from side to side – the elements were arranged from left to right in ascending order of atomic weight – and another set of relationships when read up and down – the columns grouped elements with similar valencies and properties (metals and gases, for example).

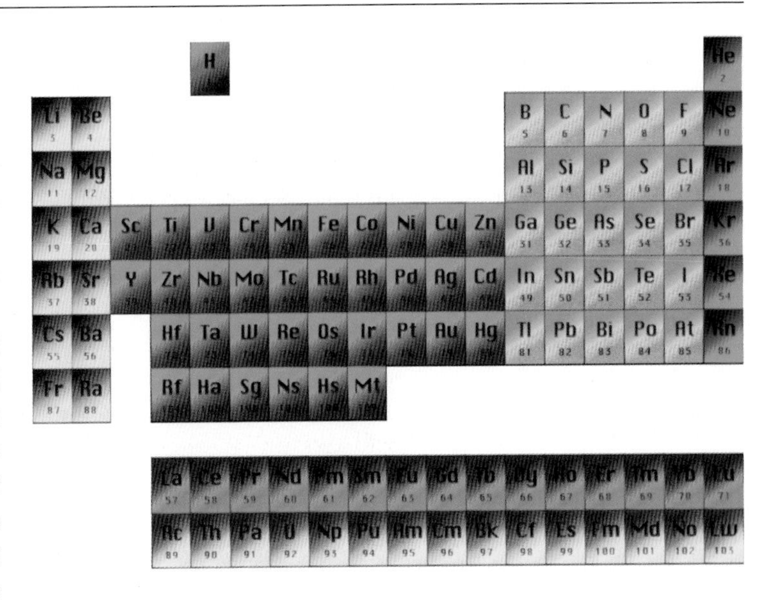

Mendeleyev's periodic table. The elements can be grouped vertically as 'families', or horizontally as 'periods'.

There had already been several attempts to organize the elements, all thus far unsuccessful. Some had tried to organize them by their properties: metals, gases and so on, but this was regarded as too simplistic. More promisingly, some chemists had tried to organize them by atomic weight. The English chemist John Dalton first developed the idea of atomic weight – the total number of subatomic particles in the atoms of a particular element – in 1803. Since then, chemists such as Johann Döbereiner and William Odling had used this to try and find numerical connections between the elements, and so organize them into different classes.

In March 1866, an English sugar refiner and amateur chemist named John Newlands presented a paper to the Chemical Society with his own idea for bringing order to the elements. He had noticed that when the elements were arranged in order of increasing atomic weight, every eighth element was related – or had properties similar to – the first element in the group. Elements, he concluded, were multiples of 8, like notes in an octave. He therefore christened his system the 'Law of Octaves'. Perhaps it was the manner of his presentation, or the fact that he was an amateur, but sadly for Newlands, his idea was greeted with general derision and mockery. One chemist asked if he could get his elements to play them a little tune. Another sarcastically commented that he might just as easily have arranged the elements alphabetically – implying that his system was based on coincidence. What none of them realized was that Newlands had come very close to discovering a useful table of the elements. His system had its faults, but given time and encouragement he could have ironed them out. However, he was so disheartened by his reception that he gave up the idea and retired from chemistry for good.

In 1867, as he worked on his *Principles of Chemistry*, Mendeleyev was unaware of the efforts of Döbereiner, Odling and Newlands. He was simply trying to solve a problem with the structure of his book. He wondered if there might be a relationship between atomic weights and the properties of elements. He tried ordering them by atomic weight, and began to notice a pattern. Like Newlands he saw that the properties periodically repeated themselves, which was how the Periodic Table eventually got its name.

Unlike Newlands, Mendeleyev placed his elements in groups of 7, but the basis of his scheme was fundamentally the same. The advantage he had was that he could back it up with a lot more data than Newlands had available to him. Furthermore, Mendeleyev had the confidence to revise the atomic weights of a few of the elements where his scheme seemed to demand it – and he was later proved right to do so. As the table developed, new and hitherto unsuspected chemical relationships between elements were revealed. By March 1869, Mendeleyev had assembled over 60 elements into his new table, and he made a formal presentation of the scheme to the Russian Chemical Society.

Predicting new elements

The new system did not win immediate acceptance. Its greatness only became apparent as time passed. The table had gaps, but Mendeleyev predicted that these would be filled by elements yet to be discovered. In November 1870, he even went so far as to describe the properties of three of these as yet unknown elements, which he named eka-aluminium, eka-boron and eka-silicon. Within 16 years, all 3 had been discovered and named respectively gallium (1875), scandium (1879) and germanium (1886) – and all three conformed very closely to Mendeleyev's predictions.

These vindications of Mendeleyev and his table won him international renown, and the Russian chemist received numerous awards and honorary degrees from universities all over the world. As time passed, new elements were discovered, and were found to fit neatly into his table. Today, the Periodic Table contains 90 naturally occurring elements, and about 24 others that have been created in laboratories.

Mendeleyev continued to teach at the university for a further 20 years after creating his table. He was an enthusiastic and popular lecturer, famous for his wild hair and beard, which he reputedly had trimmed just once a year.

Though first and foremost a scientist, Mendeleyev was also interested in politics and the state of the nation. His political activism got him into trouble with the government and the university, and he was forced to resign from his post in August 1890. However, he was such a nationally popular figure that the government felt obliged to find him other employment. In 1893 he was appointed director of the Bureau of Weights and Measures, a position he held until his death.

Dmitri Mendeleyev is today remembered as the man who discovered the inter-relationship between the chemical elements, and in so doing transformed chemistry into a logical and coherent subject for study. He was a man who lived for his work. As he said: 'Look for peace and calm in work: you will find it nowhere else. Pleasures flit by – they are only for yourself; work leaves a mark of long-lasting joy, work is for others.'

James Clerk Maxwell

Scottish physicist James Clerk Maxwell (1831–79) was arguably the greatest scientist of the nineteenth century. His work in the areas of electromagnetism, the molecular behaviour of gases, astrophysics and colour perception were unparalleled and have had a major impact on the world we now live in.

1831–79

James Clerk Maxwell was born in June 1831 in Edinburgh, Scotland, the only son of a lawyer named John Clerk. Soon after James's birth, the family relocated to a country estate at Glenlair in Kirkcudbrightshire that John Clerk had inherited from his Maxwell ancestors. At the same time, the family adopted the additional surname Maxwell. Here the young James enjoyed a comfortable rural upbringing. He received his early education from his Christian mother, but she died when he was just 8. Her influence must have been strong, however, for Maxwell remained a devout Christian throughout his life. His father at first hired a private tutor to teach the boy, but then enrolled him at Edinburgh Academy in 1841.

A brilliant mind

James's early progress at school was not spectacular. He was a shy boy, who made few friends, and received the nickname 'Dafty'. Then, at the age of 14, he surprised everyone by suddenly revealing a brilliant mind. He wrote a complex paper describing a way of drawing mathematical curves with a piece of string. His ideas were not completely new, but nevertheless showed remarkable insight for one so young. This was followed up with a succession of prize-winning work in science and mathematics.

In 1847, when he was 16, Maxwell entered Edinburgh University, where he studied natural, moral and mental philosophy. Whilst there, he published two scientific papers in the journal of the Royal Society of Edinburgh. Then, in 1850, Maxwell was accepted at Cambridge University to study mathematics. Fellow undergraduate William Thomson (later Lord Kelvin) described Maxwell in his student days: 'The impression of power which Maxwell produced on all he met was remarkable; it was often much more due to his personality than to what he said, for many found it difficult to follow him in his quick changes from one subject to another, his lively imagination started so many hares that before he had run one down he was off on another.'

On graduating in 1854 with first-class honours, Maxwell was offered a fellowship at Trinity College, Cambridge. He wrote 2 papers at this time, 'On the Transformation of Surfaces by Bending' and 'On Faraday's Lines of Force'. The latter showed how simple mathematical equations could express the relationship between the forces of electricity and magnetism. In this he took the first steps into an exploration of electromagnetism, the work for which he is best remembered.

Saturn's rings

In November 1856, Maxwell was appointed professor of natural philosophy at Marischal College, Aberdeen. Not long after his arrival there, he heard that the subject of the Adams Prize of 1857 was the motion of Saturn's rings. This was a subject that had intrigued Maxwell since his school days, and he decided to compete for the prize. He demonstrated that the stability of the rings could only be achieved if they were made up of numerous small solid particles rather than being completely solid or liquid as had been suggested by some. His essay won the prize. One of the judges, the British astronomer Sir George Biddell Airy, said of Maxwell's work: 'It is one of the most remarkable applications of mathematics to physics that I have ever seen.' Maxwell's conclusions were corroborated in 1981 when the *Voyager I* space probe took close-up photographs of the rings.

In 1860, Maxwell was appointed professor of natural philosophy at King's College London. His responsibilities here were more onerous than those at Aberdeen, with lectures for nine months of the year, as well as evening classes. Nevertheless, during the five years he spent at London University, he did the most remarkable work of his life – on electromagnetism.

Electromagnetism

Scientists had been aware since 1820 that electricity and magnetism were somehow connected, when the Danish physicist Hans Christian Oersted made a remarkable discovery in the course of a lecture. Passing an electric current through a wire on his desk, he noticed that the needle of a nearby compass was deflected from north. He realized that the electric current had produced a magnetic field around the wire. Following Oersted's discovery, the English scientist Michael Faraday wondered if the opposite might be true – could magnetism somehow produce electricity? In 1831 – the year of Maxwell's birth – Faraday proved that it could by showing that when a wire moves within the field of a magnet, it causes an electric current to flow along the wire. This effect is known as electromagnetic induction and is the principle behind the operation of electric generators and dynamos.

Einstein described Maxwell's discoveries as 'the most profound and the most fruitful that physics has experienced since the time of Newton.'

Faraday went on to develop some theories about the connection between electricity and magnetism, but could not complete his work. Maxwell took up the challenge and searched for an explanation for the relationship between the two forces. He soon realized that electricity and magnetism were simply alternative expressions of the same phenomenon – electromagnetism. He proved this by producing intersecting electric and magnetic waves from a simple electric current. Maxwell expressed this mathematically in four linked equations, now collectively known as Maxwell's equations, which he presented to the Royal Society in 1864.

Maxwell's equations showed that these electric and magnetic waves travel at a speed very close to that of light (300,000 km per second). This led him to a remarkable conclusion: that light itself was a form of electromagnetic wave; his connection of light and electromagnetism proved to be a hugely important milestone in the history of physics. Furthermore, he suggested that other types of electromagnetic waves with different wavelengths may also exist, which was verified in 1887 – eight years after Maxwell's death – when the German physicist Heinrich Hertz produced the first man-made radio waves. Further confirmation of Maxwell's theory followed with the discovery of X-rays in 1895.

The behaviour of gases

In 1865, Maxwell returned to Scotland and took up residence at Glenlair. His attention turned to the problem of the behaviour of gases. This work was a continuation

The trichromatic theory of colour

Between 1849 and 1860, Maxwell researched the concept of colour and how we perceive it. The first person to make a scientific study of colour was Isaac Newton (1642–1727), who claimed that there were seven basic colours from which any combination of colour can be produced. In 1801, Thomas Young used spinning coloured discs to show that the eye recognizes just three primary colours – red, green and violet (later modified by another scientist, David Brewster, to red, green and blue). This is known as the trichromatic theory of colour. Maxwell's big contribution to the field came in his paper 'On the Theory of Colour Vision', which was presented in 1860. In this paper, which won the Rumford Medal, he conclusively proved the accuracy of the trichromatic theory. He also showed that colour blindness is due to a person's inability to recognize red light.

In 1861, Maxwell found a practical application for the theory by creating the first-ever colour photograph. He asked the photographer Thomas Sutton to take three black and white photographs of a tartan ribbon, each time with a different colour filter – red, green and blue – placed over the lens. The three images were then projected and superimposed through the same filters, forming a full-colour image. The trichromatic process is the basis for modern colour photography.

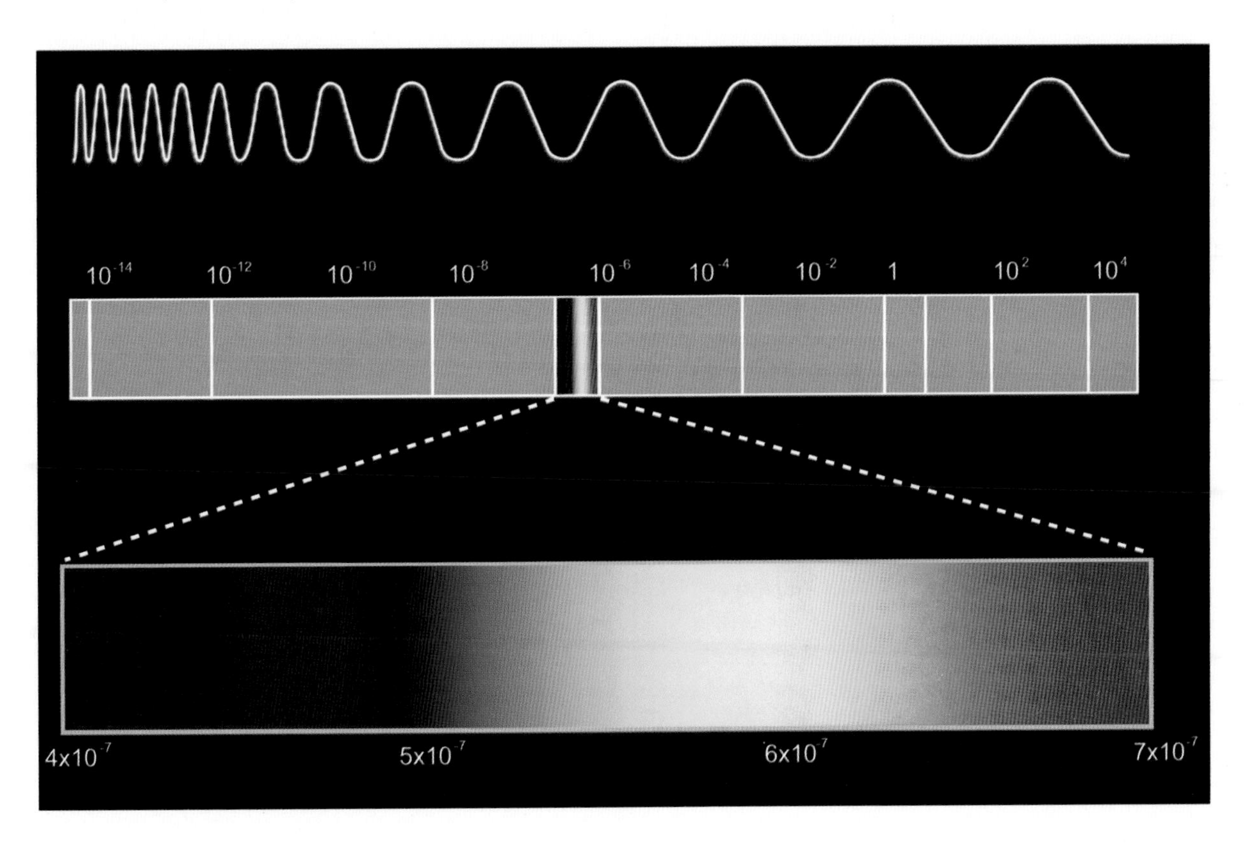

of studies by a number of earlier scientists, including the English physicist James Joule. In 1840, Joule had discovered that heat is a result of the movement of molecules. This gave rise to a new scientific discipline called thermodynamics, which includes the study of how gas molecules move. Eight years later, Joule managed to estimate the speed of gas molecules. However, he assumed that all molecules travel at the same speed, whereas in fact the speed of molecules varies greatly due to their collisions with other molecules.

Maxwell realized that it would be impossible to work out the actual speed and position of every gas molecule at every moment in time. What he could do, however, was work out the *probable* distribution (that is, the speed and position) of molecules at any given moment. This application of probability to molecular activity was revolutionary, and it offered the best explanation for the behaviour of gases that had yet been devised. Maxwell presented his theory in 1866, and it came to be known as the Maxwell–Boltzmann kinetic theory of gases. (Ludwig Boltzmann was an Austrian physicist who had independently reached the same conclusions, also in 1866.)

In 1871, Maxwell was invited to become the first Cavendish professor of physics at Cambridge University. The post was named for Sir Henry Cavendish (1731–1810), an eccentric English scientist who was famous for, among other things, his accurate estimate of the density of the earth. Taking up the offer, Maxwell designed and set up the Cavendish Laboratory, which opened in 1874 and became renowned as a centre of significant research in experimental physics. He spent the next few years editing Cavendish's extensive collection of papers and repeating many of his experiments. In doing so, Maxwell greatly contributed to the history of physics by revealing that the shy and reclusive Cavendish had been decades ahead of his time, particularly in his researches into electricity. Maxwell continued in this post until his death from abdominal cancer in November 1879, at the young age of 48.

Legacy

When one looks at the sheer breadth of Maxwell's achievements, it is difficult to imagine that they were all the work of one man. His work in such diverse fields as electromagnetism, the molecular behaviour of gases, colour theory and astrophysics were all truly ground-breaking, and enabled the development of many of today's technologies. Of all these achievements, his contribution to electromagnetism was undoubtedly his greatest, and his book *Electricity and Magnetism* (1873) remains a classic work of science. As physicist Richard Feynman noted, 'From a long view of the history of mankind – seen from, say, ten thousand years from now – there can be little doubt that the most significant event of the nineteenth century will be judged as Maxwell's discovery of the laws of electrodynamics.'

Maxwell's name is not as instantly familiar as those of Newton and Einstein, partly because he was no longer alive when the significance of his work became evident, yet many regard his work as being on a par with theirs. A very modest man who did not seek public fame, Maxwell achieved happiness simply from his work. As Maxwell himself said in 1860: 'We, while following out the discoveries of the teachers of science, must experience in some degree the same desire to know and the same joy in arriving at knowledge which encouraged and animated them.'

Max Planck

Max Planck (1858–1947) was a theoretical physicist whose curiosity about the laws of nature led him to stumble on a whole new realm of physics – quantum theory – that would transform our understanding of the microphysical world of atoms and subatomic particles.

1858–1947

Max Planck was born on 23 April 1858 in Kiel, Germany, the son of Julius Wilhelm and Emma Planck. Julius was a distinguished professor of constitutional law and he instilled in his children a love of scholarship and a respect for the institutions of state and church. He also taught them the values of honesty, fairness and generosity.

Education

Max began his education at a local school in Kiel. In 1867, when Max was nine, Julius was appointed professor at the University of Munich. Max was sent to the city's famous Maximilian Gymnasium, where he did well but not brilliantly in all the subjects he studied, usually coming in the top eight in each class. Perhaps surprisingly, he did not display an outstanding aptitude for science and maths at this stage in his life. If anything, he showed more promise in music. He was blessed with the gift of perfect pitch, and was a talented pianist and organist. It was towards the end of his schooling that he began to display a deeper interest in physics and mathematics, thanks mainly to his teacher Hermann Müller. Through Müller, he learned about the law of the conservation of energy – the first law of thermodynamics. He was impressed by the idea of the world being governed by absolute laws of nature, and he wondered if there were more to be discovered.

Nevertheless, when he passed his school-leaving exam in July 1874, aged 16, Max still had no clear idea of what he wanted to do with his life. The three options seemed to be music, mathematics or physics. He discussed the possibility of a musical career with a musician who told him that if he needed to ask the question, he'd better do something else.

Max decided to study mathematics and physics at the University of Munich. He talked to his physics professor, Philipp von Jolly, about whether to devote his life to physics. The professor urged him not to, saying there were no more breakthroughs to be made in that subject. This was actually a fairly widespread feeling towards the end of the nineteenth century. Most scientists believed that the mysteries of the physical world – including the laws governing motion, gravity, electricity and magnetism, gases, optics and many other things – had been revealed. They predicted that physics in the coming century would be more about consolidation and refinement of existing knowledge.

Despite von Jolly's off-putting remarks, Planck decided to become a theoretical physicist. His decision was based on his fascination with the laws of the universe and his belief that the gift of reason enabled human beings to gain insight into the workings of the world. In 1877, he completed his education at the University of Berlin, where he attended inspiring lectures by Hermann von Helmholtz and Gustav Kirchhoff, both leading physicists of their generation. Planck was also greatly influenced by his study of Rudolf Clausius's articles, which introduced him to the significance of the second law of thermodynamics (see box on p. 114).

Planck returned to Munich and obtained his doctoral degree in July 1879 at the age of just 21 with a thesis on the second law of thermodynamics. He completed his qualifying dissertation the following year and became a lecturer at Munich. In 1885, aged 27, he was appointed associate professor of theoretical physics at the University of Kiel. This job gave him sufficient financial security to move out of home and marry Marie Merck, a Munich banker's daughter whom he had known and loved for many years. They married on 31 March 1887.

In October of that year, Kirchhoff, Planck's former teacher at the University of Berlin, died. The university wished to fill the post with a world-renowned physicist. Their first choices – Ludwig Boltzmann and Heinrich Hertz – turned them down, clearing the way for Max Planck. He took up the post in 1888, and in 1892 he was promoted to full professor there – a position he retained until his retirement in 1927.

'The outside world is something absolute. And the quest for [its] laws appeared to me as the most sublime scientific pursuit in life.'

While at Berlin, Planck did his most brilliant work in theoretical physics. He began by continuing the work in his dissertation on the second law of thermodynamics and the concept of entropy. He investigated how materials transform between solid, liquid and gaseous states. He also looked at the conduction of electricity through liquid solutions (electrolysis). In doing so he managed to find explanations for the laws governing the differing freezing and boiling points of various solutions.

Research into radiation

In the mid-1890s, Planck turned his attention to the question of how heated substances radiate energy. Physicists were aware that all bodies radiate heat at all frequencies – although maximum radiation is emitted only at a certain frequency, which depends on the temperature of the body. The hotter the body, the higher the frequency for maximum radiation. (Frequency is the rate per second of a wave of any form of radiation.) Planck wanted to see if this process was governed by a universal law. It was difficult to obtain accurate measurements of things like radiation and frequency in the laboratory because hot bodies behave irregularly, and so Planck made use of a 'blackbody', a hypothetical object that completely absorbs and then re-emits all radiation falling upon it. He began analysing the 'spectral energy distribution' of the blackbody – the curve displaying how much radiation it emitted at different frequencies for a given temperature.

In 1896, William Wien, a member of staff at the Physikalisch-Technische Reichsanstalt (PTR), a centre of radiation research in Berlin, suggested a formula that seemed to fit the spectral energy distribution of the bodies he had experimented on. Over the next few years, Planck made a series of attempts to make this formula fit with his own theoretical experiments on the blackbody. In 1900, he succeeded in doing so. However, precision measurements taken at the PTR showed that while Wien's formula was valid at high frequencies, it broke down completely at low frequencies. At the same time, two English physicists, Lord Rayleigh and James Jeans arrived at another formula which worked at low frequencies, but not at high frequencies.

Planck learned of this discrepancy in October 1900, and went straight to work on the problem. Here were two formulae, both of which worked, but at different frequencies. The solution, he guessed, must be to combine the two formulae into a single formula that worked at all frequencies. Planck very quickly found his formula: the spectral energy distribution of a blackbody could be expressed as a straightforward multiplication of frequency by a certain number, which became known as 'Planck's constant' (6.6256×10^{-34}).

The new formula, known as Planck's radiation law, was acclaimed by fellow physicists as undoubtedly correct. To Planck, however, it was simply a 'lucky intuition' – something designed to fit experimental results. If it was to be taken seriously, it needed a theoretical foundation. After less than two months of concentrated work, he succeeded in providing this. He presented his report at a meeting of the German Physical Society on 14 December 1900.

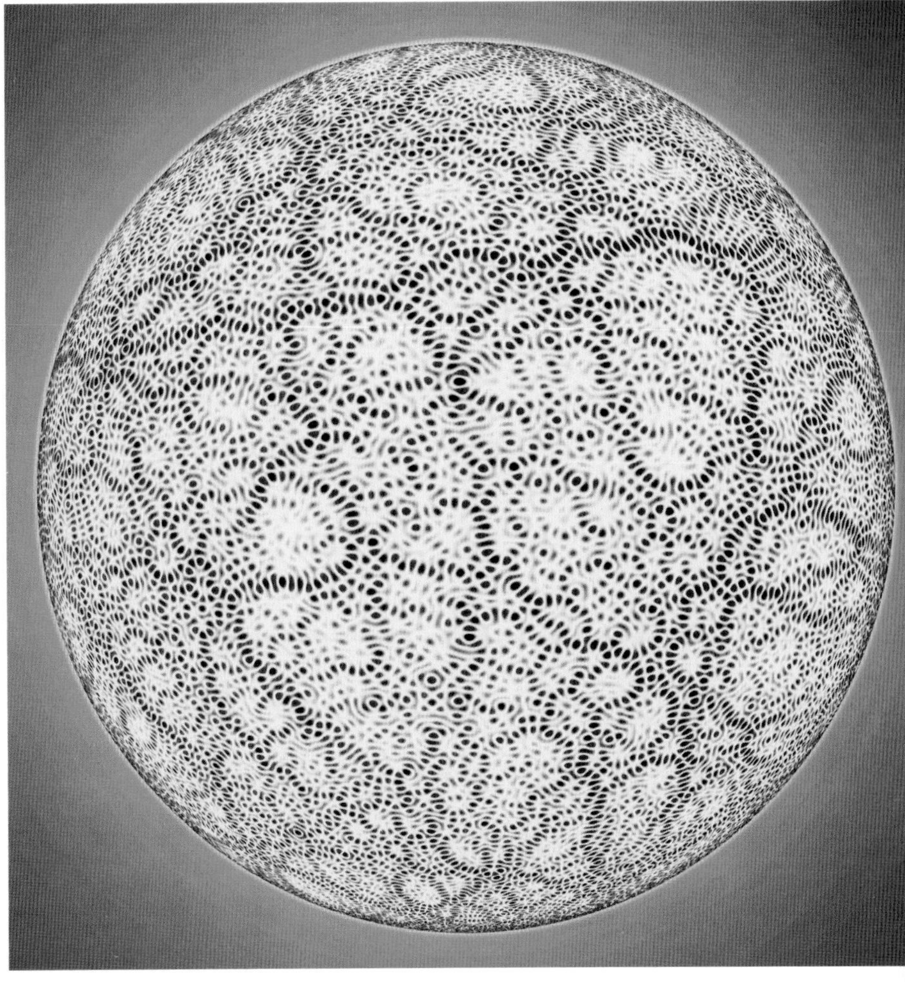

Computer model showing many wave paths superimposed onto the surface of a sphere. This produces a random wave, an example of quantum chaos. Classical chaos is when waves can travel in any possible direction. Quantum chaos is when quantum wavepackets are added together randomly. This model of a random wave was produced as an artwork by Professor Eric Heller.

Quantum theory

There was a problem, however. Planck's new radiation law completely contradicted a basic assumption of physics about the nature of energy. According to his law, the energy released (e) by a hot body is equal to the frequency of the radiation (n) multiplied by Planck's constant (h). This can be expressed in the formula $e = hn$. Now, h is a tiny number, close to zero, yet it has a finite value. In other words, $e = hn$ seemed to imply that energy was released in tiny, indivisible chunks, which Planck called 'quanta'. This was a revolutionary notion. Until that point, it had always been assumed that energy was released in a continuous stream. Physicists had suddenly to get used to the idea that the microphysical world – the world of atoms – could not be described in terms of classical physics. By introducing the idea of quanta, Planck had opened the door to a new kind of physics – quantum theory.

Neither Planck nor his contemporaries immediately understood the revolutionary nature of quantum theory. It did not start to become apparent until 1905, when

The second law of thermodynamics

Max Planck was fascinated by the second law of thermodynamics. This law, formulated by Rudolf Clausius in 1850, deals with the concept of entropy, which is a measure of the 'disorder' in a system. In simple terms, the second law states that in all natural processes the entropy of the world always increases. So, for example, if a cup of hot tea is placed in a cold room, there is an uneven distribution of heat (or energy) in the room; this causes a flow of energy from the cup to the room until tea and room are the same temperature. In other words, a force has acted to minimize the disequilibrium of energy and maximize entropy.

Albert Einstein proved that light – another form of energy – was also emitted in quanta, later called photons. Gradually, other scientists began doing work in this area. The French mathematician Jules Poincaré proved mathematically that quanta were indispensable to Planck's radiation law, while the Danish physicist Niels Bohr, with his quantum theory of the hydrogen atom, showed that Planck's constant provided the key to understanding atomic processes. By 1911, quantum theory had moved to the forefront of physics research. In the 1920s, it provided the basis for quantum mechanics, which explored the properties of atoms and molecules.

Max Planck was already 42 when he unveiled quantum theory, an age when most physicists have already done their best work, and so further developments in the field were left to younger minds. Ironically, he was one of the last to accept the full implications of the theory he had set in motion. He loved the ordered logical world of classical physics and was reluctant to abandon it. Quantum theory, by contrast, seemed to suggest a microphysical universe of paradox and uncertainty. Light, for example, could behave either as waves or as particles (photons), depending on the type of experiment performed.

Later life and legacy

After 1900, Planck continued to contribute research at a high level in various fields, including optics, thermodynamics and physical chemistry. In 1918 he was awarded the Nobel Prize for Physics for his development of quantum theory. His later life, however, was marked by a series of personal tragedies. In 1909, his wife Marie died, leaving Planck with two sons, Karl and Erwin, and twin daughters, Margrete and Emma. Karl was killed in action in World War I in 1916. The following year, Margrete died in childbirth. In 1919, Emma died in the same circumstances. In 1944, Planck's Berlin house was hit by allied bombs and many of his scientific papers were destroyed. That year, Erwin was accused of being part of the plot to assassinate Hitler and executed by the Gestapo. After the war, Planck moved to Göttingen. He died there in 1947, aged 89.

Today, Max Planck is primarily remembered as the man who, through quantum theory, revolutionized our understanding of the microphysical world. However, his achievements did not end there. He was an outstanding teacher. His five-volume *Introduction to Theoretical Physics*, a collection of his lectures published in the 1920s, is regarded as a classic. As professor of theoretical physics at Berlin, he raised the status of the subject immensely and supported the work of many younger physicists, including Einstein. Science is truly fortunate that Max Planck chose to disregard the advice of his teacher, Philipp von Jolly, given in 1874: 'Theoretical physics is a very fine subject, of course, ... but it is unlikely that you will be able to add anything new of fundamental importance to it.'

Marie Curie

Marie Curie (1867–1934) became the first person to isolate the radioactive elements polonium and radium, earning her the 1903 Nobel Prize for Physics, together with her husband and colleague, Pierre. Marie Curie was also awarded the Nobel Prize for Chemistry in 1911.

1867–1934

TODAY, THE VERY WORD 'RADIOACTIVITY' USUALLY STRIKES FEAR into the best of us, having the associations it does with nuclear warfare, radioactive poisoning and cancer. Knowing nothing of its lethal properties, Marie Curie embarked upon dogged research into the topic. Her progress is all the more remarkable when one considers the difficulties of her career as a scientist at a time when society frowned upon women working in this field, and her determination and courage in pursuing her discoveries in a life dogged by financial difficulties, ill health and personal tragedy.

Early years

Marie was born in Warsaw, Poland, on 7 November 1867 and christened Marya Sklodowska (she changed her name to Marie when she moved to France). Her parents were both teachers, but it was Marie's father who inspired in his daughter a love of science and nature. Marie's lifelong struggle with personal tragedy began when she was seven and her elder sister Zofia died from typhus. Her mother died just 4 years later from tuberculosis. Marie became a serious, studious child who regularly topped the class. By the time she was 18, she was set upon going to university to study physics. As a woman she was barred from higher education in her native Poland, and so she planned to go to the Sorbonne in Paris, France. The family had little money, and she struck a deal with her sister Bronia to get around the situation. Marie would work as a governess to fund her sister's passage through university, and then, when Bronia was qualified and working, she would pay for Marie's education. Marie spent 8 long frustrating years as a governess until November 1891, when she finally boarded the train to Paris.

Paris, at last!

Not only was Paris an exciting city, buzzing with cultural and artistic vitality, but for Marie it meant freedom to devote herself to learning all there was to know in the field of physics. The laboratories at the Sorbonne were well equipped and run by some of the most respected scientists of the day. Though Marie was just one of a handful of women studying science at the great university, she felt at home. She dedicated her life to her studies and in 1893 graduated in physics, taking first place in the year. Two years later, Marie completed a second degree, in mathematics.

In 1894 Marie met Pierre Curie (1859–1906). The quiet, grave scientist, a professor at the School of Physics at the Sorbonne, was Marie's perfect other half. Though dedicated to his profession, he was also at home with nature and loved being outdoors. The couple were married in 1895 and lived frugally in a small flat in Paris. In 1897 Marie gave birth to their first daughter Irène. That same year Marie began her PhD dissertation, an investigation into the properties of the element uranium.

X-rays and beyond

In 1895 a German physicist called Wilhelm Röntgen (1845–1923) had discovered the electromagnetic rays that became known as X-rays. Then, in 1896, the French physicist Antoine Henri Becquerel (1852–1908) found that uranium salt produced what he thought were similar rays, which were subsequently called 'Becquerel rays'. He deduced that these rays were a property of atoms. The discovery of X-rays and Becquerel rays opened up a whole new field of research for scientists and in turn started the nuclear age.

Marie began her own research to determine which elements gave off similar 'emissions'. She set up her laboratory in a dark and dusty storeroom at the Sorbonne School of Physics. With no money to fund her research, she made use of her husband Pierre's measuring instruments. In 1898 Marie coined the term 'radioactive' to describe the elements that gave off the mystery rays. Eventually, Marie found that the compound pitchblende produced more radioactivity than uranium, a fact that led her to conclude that it contained a more radioactive element than uranium. As yet this element was undiscovered, which presented Marie with the challenge of a lifetime: she needed to isolate the element to prove to the world that it actually existed. The problem was so tantalizing that Pierre joined his wife in her quest.

'Nothing in life is to be feared. It is only to be understood.'

Dirty work

Great sacks of pitchblende were delivered to the cramped laboratory, where there was little room for manoeuvre. The processing of the pitchblende was run like a military operation. First it was was ground down a kilogramme at a time, and then it was sieved before being boiled and continually stirred for hours to form a liquid which could be distilled. Finally, the liquid was electrolyzed until a minute amount of the radioactive element was isolated. The process was backbreaking, but Marie battled on, determined to find the element that she would name after her homeland. In spring 1898 Marie and Pierre discovered polonium. It was an extraordinary element that glowed in the dark when it was mixed with water.

The Curies' triumph was short-lived, for they quickly deduced that pitchblende contained another even more radioactive element. More tiring days and nights bent over bubbling containers of the radioactive broth ensued. It took the determined pair four more years to extract a fraction of a gramme of radium. During those years they lived on a shoestring. At this time, nobody realized the danger of working so closely with radioactive materials, and the aches and pains and weight loss they were suffering were explained away as the results of hard work and poor diet. A few months before their great discovery, Marie also lost her beloved father.

A Nobel Prize for a woman!

In 1903 the Swedish Academy in Stockholm jointly awarded the Curies and Antoine Henri Becquerel the Nobel Prize for Physics for that year. It was a wonderful honour but Marie had nearly missed out on the prize altogether, because many of the judges had been against awarding the prize to a woman and proposed giving it to Pierre. As it happened, neither of the Curies were well enough to attend the award ceremony in Sweden. Their work with radioactive elements was having an insidious effect on their health and in August 1903 Marie had suffered a miscarriage which took months to get over.

The Nobel Prize brought Marie and Pierre fame throughout the world. The shy couple were suddenly thrust into the limelight; a position that neither of them took to very well. The prize did however bring them more money for their research, and they were able to to upgrade their working conditions and employ a laboratory assistant. In time it also meant that Pierre was deemed worthy enough to be appointed as professor of physics at the Sorbonne. At last the Curies were working in well-appointed laboratories at the university. At the end of 1904 Marie gave birth to another girl they named Eve. The following year they made the long-postponed trip to Sweden to pick up their Nobel Prize. In his acceptance speech, Pierre warned of the dangers of radium. He foresaw the dangers of the radioactive element getting into the wrong hands and being used as a means of destruction.

The journals of Marie and Pierre Curie. Even now too radioactive to be safely handled, they are kept in a sealed display at the Bibliothéque Nationale, Paris.

Beyond grief

'Marie Curie is dead to the world. She is a scientist walled behind her grief.' Marie's friend Marguerite Borel wrote these words in 1910. In 1906 Pierre had been tragically run over by a horse and cart and killed. His sudden death plunged Marie into despair; her only solace being her children and her work. Soon after her husband's death she was offered his position as professor of physics at the Sorbonne. Marie had reservations about accepting the role but after due consideration realized it was the best thing she could do to continue her work. At the same time she was making history by becoming the first female professor at the university.

Radioactivity was Marie's life's work. In 1911 she was awarded the Nobel Prize for Chemistry for her work on polonium and radium. Then, in 1913, she established a research laboratory for radioactivity which became known as the Paris Radium Institute. Marie was closely involved in the design of the building, ensuring that the laboratories were large and airy and properly equipped. Marie's own research was taking her closer to the reason why some elements were radioactive. Though Marie never found the answer to this question, she watched with interest as such scientists as the New Zealand physicist Ernest Rutherford (1871–1937) unlocked the secrets of the atom, which in turn led to a greater understanding of radioactivity (see box).

'Little Curies'

When the First World War began in August 1914, Marie lost little time in removing the supply of radium from her laboratory to a bank vault in Bordeaux. Marie considered the supply so precious, and deemed it to be so dangerous in the hands of the wrong person, that she undertook the transportation of the chemical herself. Later in the war she joined forces with her daughter Irène to take X-ray equipment out onto the battlefields of France where it could be used to treat the injured soldiers. She raised the funds for a special vehicle (designed by herself) that could carry the X-ray machinery to where it was needed. By the end of the war, mother and daughter had kitted out 18 'Little Curies', as these vans became known. Marie was also personally involved in the setting up over 200 stationary X-ray clinics. Though she suffered terribly with her own health, Marie personally operated the equipment and was involved in training other women to use it.

Precious elements

Following the war, Marie was appointed director of the Paris Radium Institute, a position she held until her death in 1934. The year after her appointment she was awarded an honorary professorship of radiology by the Warsaw Radium Institute, a title that meant a great deal to her personally because it was her home town.

In 1920 Marie was visited by the American journalist Marie Meloney, who was stunned by the lack of funding for Marie's work. At this time there was only one gramme of radium in existence, housed at the Paris Radium Institute. Meloney promised Marie that she would raise the money in America to buy another. In 1921 Marie made her first trip to the United States where she was welcomed at the White House by President Warren Harding and presented with a casket containing the precious element. She made a second visit in 1929 to receive radium for the Warsaw Radium Institute.

The final years

Throughout her later years Marie constantly battled with illnesses, and her eyesight deteriorated badly, but she remained as indefatigable as ever when it came to her work. She expanded the collection of radioactive substances at the institutes and was personally involved in the recruitment and training of hundreds of new scientists. In 1934 she was finally diagnosed with leukaemia, a life-threatening disorder that we now realize can be caused by contact with radioactive materials. Marie died on 4 July 1934 at the age of 66. In 1995, at the request of President François Mitterand, Marie and Pierre Curie's remains were transferred to the Panthéon in Paris, the monument and resting place for France's greatest heroes.

The atom

In the early nineteenth century, scientists believed that the atom was a single indivisible particle, but by the end of the century they had discovered that atoms contained electrons. It was the work of Ernest Rutherford that brought the first real understanding of the atom. Following his experiments with helium atoms, he concluded that the atom was like a miniature universe with its mass concentrated in the nucleus. Scientists went on to conclude that the nucleus contained tiny particles called protons and neutrons. Subsequently they discovered that energy could be released by splitting the nucleus. This release of energy could be used to make electricity or to create huge explosions. They also found that they could not split the nucleus of all elements, and in fact it was only possible with radioactive elements.

Ernest Rutherford

When Becquerel discovered radioactivity in 1896, it prompted a number of scientists to probe more deeply into the phenomenon. Among those who took up the challenge was a young physicist by the name of Ernest Rutherford (1871–1937).

It was Rutherford who did more than anyone else to reveal the true nature of radioactivity. He also made significant advances in understanding the architecture of the atom and laid the foundations for the new discipline of nuclear physics.

Early life

Rutherford was born in Spring Grove in rural Nelson, New Zealand, in 1871. He was the fourth of twelve children born to James and Martha Rutherford. At the age of 16, he won a scholarship to Nelson College. He excelled academically and at 19, won another scholarship to study at Canterbury College, Christchurch. He graduated in 1892 and got his MA the following year in mathematics and physics.

Rutherford stayed on for another year to do further research in physics. He wanted to study the magnetic properties of iron when it was exposed to high-frequency alternating electric currents. To help him with his experiments, he built a special mechanism capable of measuring time intervals of a hundred thousandth of a second. His two impressive papers on this work were sufficient to win him an '1851 Exhibition' scholarship, which provided the means for him to continue his education in England.

In 1895, aged 23, Rutherford left New Zealand for the University of Cambridge. He already had three university degrees and a growing reputation as a brilliant experimental researcher in electricity and magnetism. He decided to work with Professor J. J. Thomson at the university's Cavendish Laboratory. Here he invented a mechanism that could detect electromagnetic waves over a distance of few metres, and even through a brick wall. Rutherford worked on the sensitivity of the instrument until, in February 1896, he managed to detect electromagnetic waves over a distance of several hundred metres – a world record at that time.

Professor Thomson, realizing that Rutherford possessed a rare talent for research, invited him in early 1896 to take part in a new study. The previous December, the German physicist Wilhelm Röntgen had reported his discovery of X-rays, and Thomson asked Rutherford to help him investigate the effects of passing a beam of X-rays through a gas. They discovered that X-rays produce great quantities of ions – atoms that have acquired either a positive or negative electric charge by losing one or more electrons – and that these ions then recombine to form neutral molecules. Working alone, Rutherford then worked out a method of measuring the speed at which the ions recombine.

Investigating radioactivity

That same year, in Paris, the French physicist Antoine-Henri Becquerel made a startling discovery. He had left a packet of uranium salts in a drawer on top of a photographic plate, and later found that the salts had fogged the plate. It was clear that they were emitting radiation of some kind. In 1898, Marie Curie, a student of Becquerel's, discovered, with her husband Pierre, that other kinds of element also emitted radiation. She coined the term 'radioactivity' to describe the phenomenon.

Rutherford decided to investigate. He found that, like X-rays, radioactivity also produced ions when passed through the air. However, unlike X-rays, it consisted of two distinct types of ray. The first type, which Rutherford named alpha rays, produced very large quantities of ions but were easily absorbed by a surface. The second type, which he named beta rays, produced fewer ions, but were much more penetrative, and could pass through aluminium foil one-fiftieth of a millimetre thick. Rutherford correctly surmised that the rays were actually composed of minute particles.

In 1898, Rutherford accepted the post of professor of physics at McGill University in Montreal, Canada. Here, Rutherford continued investigating radioactivity. Together with a young chemist named Frederick Soddy, he studied the phenomenon in three elements: thorium, radium and actinium. Rutherford and Soddy noticed that thorium disintegrated into a gas, which in turn disintegrated into an unknown new element that was extremely radioactive. The radioactivity eventually made the new element disappear.

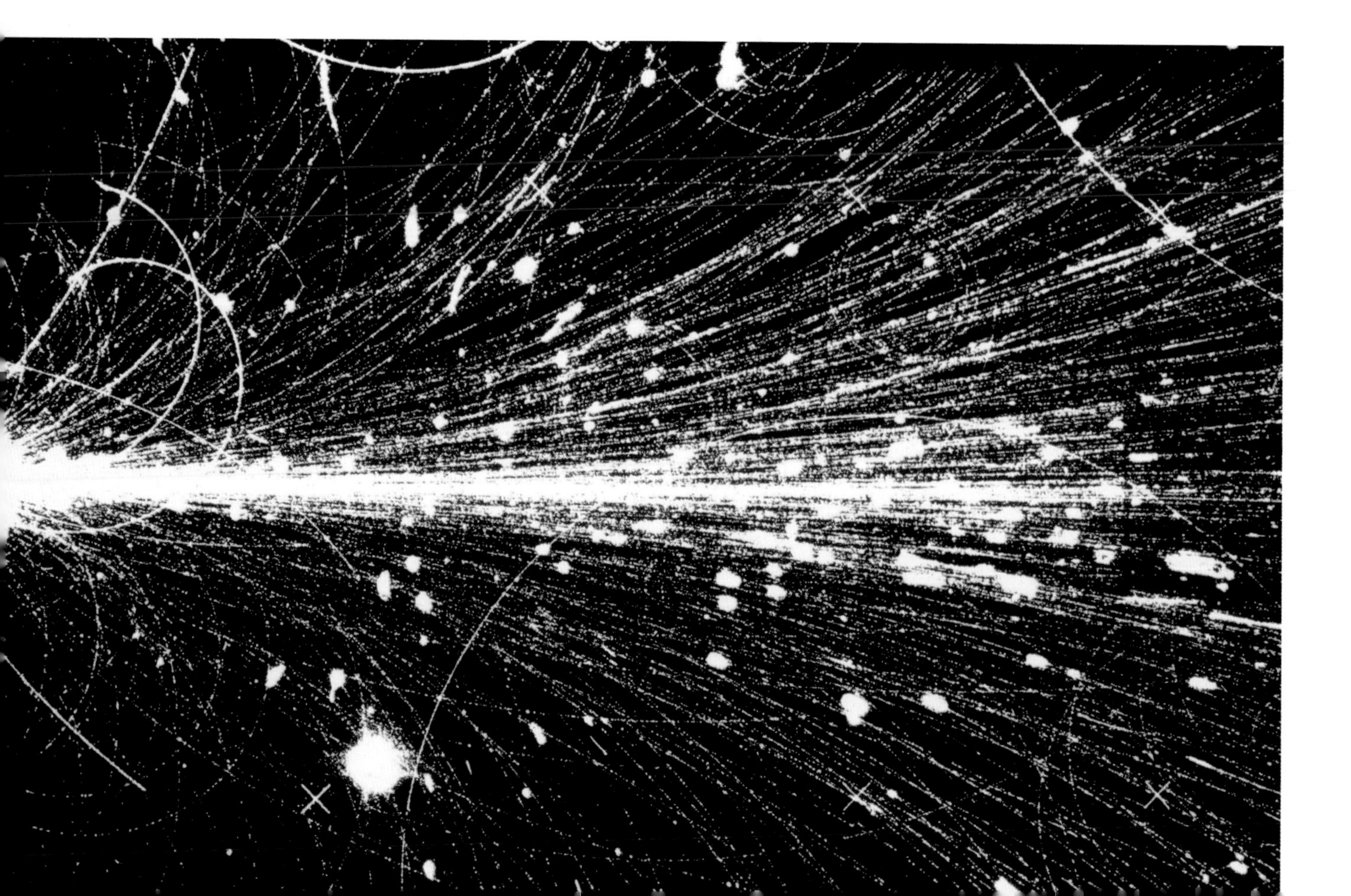

A high-energy collision of a sulphur ion with the nucleus of a gold atom at CERN, the European particle physics laboratory outside Geneva.

In 1902, Rutherford and Soddy concluded that radioactivity was a process in which the atoms of one element spontaneously changed into atoms of a different element, which was also radioactive. Until this time, scientists had firmly believed that atoms were the unchanging building blocks of nature. The idea that atoms could rip themselves apart and change into different kinds of atoms – in other words, that one element could change into another – smacked of medieval alchemy and was firmly resisted by many chemists.

Despite the controversial nature of his discoveries, Rutherford was elected to the Royal Society in 1903 and was awarded the Rumford medal in 1904 in recognition of his achievements. In 1908, as further experimentation proved the correctness of Rutherford and Soddy's conclusions, Rutherford was awarded the Nobel Prize for Chemistry.

A theory of the atom

The year before this, he had returned to England to take up the position of professor of physics at Manchester University. Here he continued his research into alpha rays, now known to be composed of particles. He and his assistant Hans Geiger constructed a mechanism that allowed them to count the particles one by one as they emerged from a known quantity of radium. By measuring the total charge collected, they were able to deduce that alpha particles were positively charged. In 1908, Rutherford allowed alpha particles to pass through a thin glass wall of a container into an outer glass tube, and found that the collected gas was helium, thereby proving that alpha particles were in fact ionized helium atoms – or helium atoms stripped of their electrons.

While he was at McGill, Rutherford had experimented with firing alpha particles at a photographic plate. He had noticed that, while the image this produced was sharp, if he passed the alpha particles through thin plates of mica, the resulting image on the photographic plate was diffuse. The particles were clearly being deflected through small angles as they passed close to the atoms of mica. In 1911, Rutherford repeated the experiment, this time with a thin sheet of gold. He suggested to his assistants, Hans Geiger and Ernest Marsden, that it would be interesting to check whether any of the particles were scattered backwards – that is, deflected through an angle of more than 90 degrees. To their astonishment, they discovered that a small proportion of the particles did indeed scatter through more than 90 degrees, emerging from the same side of the gold foil as they had entered. Rutherford described it later as 'quite the most incredible event that has ever happened to me in my life. It was almost ... as if you had fired a fifteen-inch shell at a piece of tissue paper and it came back and hit you.'

'It was almost as if you had fired a fifteen-inch shell at a piece of tissue paper and it came back and hit you.'

After doing some calculations, Rutherford concluded that only a powerful positive charge at the heart of the atom could have caused these positively charged particles to be deflected so dramatically. And because only about one in 8,000 particles bounced back, this 'nucleus' must be approximately 8,000 times smaller than the entire atom. The rest of the atom must be 'empty space', allowing the other particles to pass through. The positive charge on the nucleus was clearly balanced by an equal negative charge on the electrons distributed around it. Rutherford's theory of the structure of the atom was not wholly new. In 1904, Japanese physicist Hantaro

Discovering the half-life

During their experiments on radioactive elements between 1901 and 1903, Rutherford and Frederick Soddy found that these substances decayed at a regular rate: it always took the same amount of time for half the atoms in a sample to decay. For example, if half of a thorium sample decayed in four days, then half of the remaining half would decay over the following four days, and so on. Rutherford saw a practical use for this. He realized that this 'half-life'– this steady rate of decay – could be used as a means of measuring the age of a piece of rock. In order to work out how old it was, it was only necessary to measure the amount of radiation it contained and how swiftly it was decaying. He tested a piece of pitchblende (a mineral containing uranium) and found it to be 700 million years old – proving that the earth was far older than most experts at that time believed. As a by-product of his work, Rutherford had made a significant discovery in an entirely different field, and pioneered a new science – radiometric dating.

Nagaoka had proposed a model with electrons rotating around a central nucleus. But Nagaoka's model had been rejected because, according to the classical laws of physics, the orbiting electrons would almost immediately lose their energy and fall into the nucleus. Now, however, Rutherford had provided experimental evidence that Nagaoka might have been right after all. In 1913, the Danish physicist Niels Bohr showed that electrons – contrary to the classical laws of physics – do not lose their energy during rotation and do in fact occupy certain well-defined orbits around the nucleus, thus confirming the views of Rutherford and Nagaoka. This theory of the structure of the atom is now known as the Rutherford atomic model.

Protons and neutrons

Rutherford received a knighthood in 1914, aged 43, in recognition of his many achievements. But he was not a man to rest on his laurels. During the First World War, he developed a system for detecting submarines by underwater acoustics. In 1919 he became the first person to disintegrate an atom artificially by causing it to collide with an alpha particle. By this means he managed to change a nitrogen atom into an atom of oxygen and an atom of hydrogen. The following year, Rutherford worked out that the collision had caused positively charged particles to be ejected from the nucleus, thus changing the nature of the atom. He named these particles protons. He then speculated on how positively charged protons could coexist in the nucleus without blowing themselves and the rest of the atom apart. He suggested that the protons must be balanced by some neutrally charged particles, which he called neutrons. The existence of neutrons was later proven by James Chadwick at the Cavendish Laboratory.

In 1919, Rutherford succeeded J. J. Thomson as the head of the Cavendish Laboratory at Cambridge. In this role, Rutherford had less time for experimental research, but he continued to give lectures and support his students. During his later years, Rutherford was the recipient of many medals, awards and honorary doctorates. He died in Cambridge on 19 October 1937 aged 66, and his ashes were buried in the nave of Westminster Abbey.

Ernest Rutherford was fortunate in many ways to be alive at a time when so many discoveries were being made in physics and chemistry. But what distinguished him from his colleagues – and the reason why he is remembered today as one of the great experimental physicists of all time – were his special qualities as a scientist. Rutherford was an extremely stubborn and patient man, willing to work at problems far longer than most. He was also blessed with an open mind, and was willing to entertain and speculate on possibilities that more traditionally minded scientists might have rejected. As his student James Chadwick said: 'Rutherford's ultimate distinction was his genius to be astonished.'

Albert Einstein

In 1905, a paper was published in a German scientific journal describing a Special Theory of Relativity. The implications of this theory were so profound they would overturn classical physics and transform the scientific view of such things as space, time, matter, energy and light.

1879–1955

THE AUTHOR OF THE PAPER WAS A 26-YEAR-OLD CLERK at a Swiss patent office named Albert Einstein (1879–1955). Einstein held no university post and had no access to a laboratory or academic library. His ideas seemed to come out of nowhere. As the physicist C.P. Snow has written, it was as if he 'had reached the conclusions by pure thought, unaided'. Ten years later, Einstein completed this one-man scientific revolution with his General Theory of Relativity, which offered a new explanation for gravity. Physics would never be the same again.

Early life

Albert Einstein was born in Ulm, Germany, on 14 March 1879, and he grew up in Munich. There was little in Einstein's early life to suggest he was destined for greatness. It is said that he did not learn to speak until he was 3. The young Albert hated school, with its tough discipline and rigid teaching methods, and showed little academic promise. His only pleasures were the violin, which he would play all his life, and mathematics. He left school at 15 without a diploma.

To avoid military conscription, Einstein gave up his German citizenship and moved to Switzerland. In Zurich, he succeeded in gaining a place (on his second attempt) at the Polytechnic to study physics and mathematics. On graduating in 1900, Einstein began working as a temporary mathematics teacher, but hoped to get to university to continue his studies. He applied for entrance to several institutions during 1900 and 1901 without success.

In 1902, Einstein got a job as a technical examiner third class at a patent office in Bern. The job gave him sufficient financial security to marry his Hungarian fiancée, Mileva Maric. It also afforded him some spare time, which he spent exercising what he described as his 'disposition for abstract and mathematical thought'. He began contributing papers to a German physics journal, *Annals of Physics*.

Miracle year

In 1905, Einstein seemed to find a new level of creativity. He submitted five papers to *Annals* in that year, all remarkable and insightful pieces of work, and one of them, truly historic. The first paper offered an explanation for the photoelectric effect (see box on p. 126). In 1921 he would receive a Nobel Prize for this paper. The second

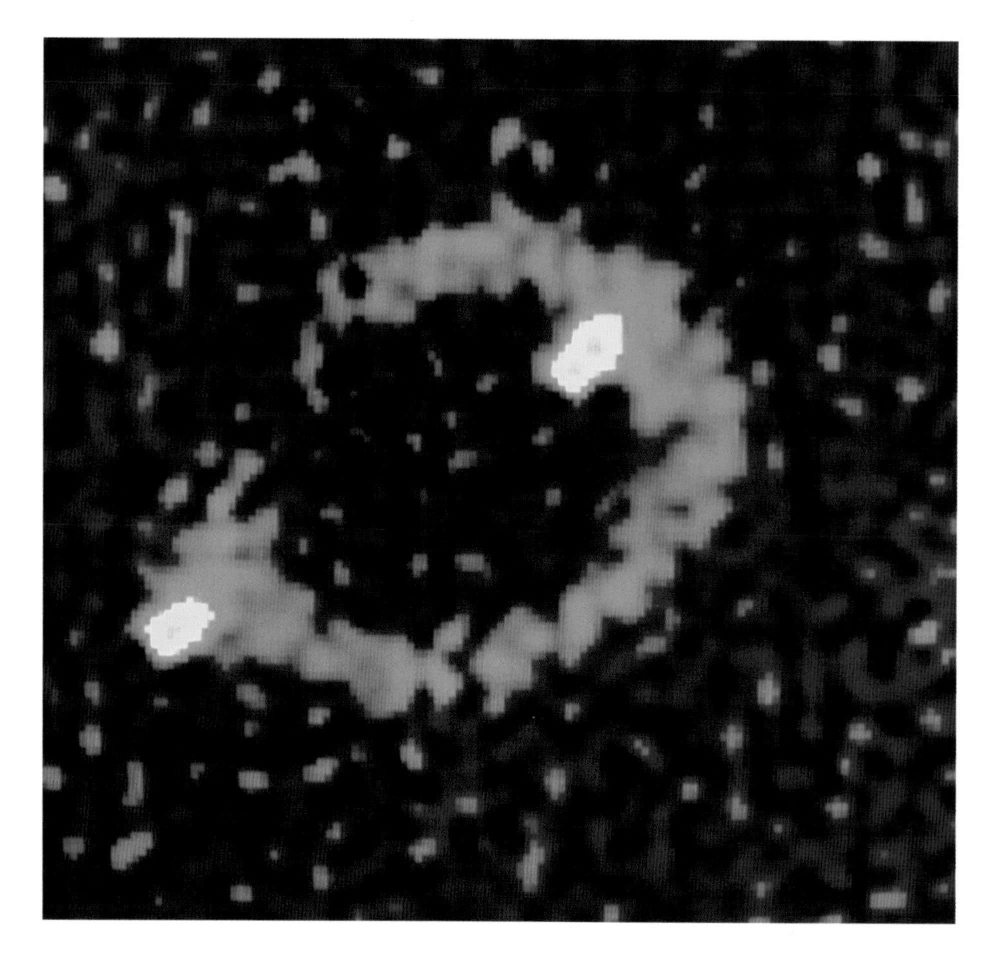

A photograph of a distant object in space known as an 'Einstein ring', first discovered in 1987. It has been suggested that it could be a distant galaxy which appears as a ring because its light has been bent, or distorted, by the gravity of an intervening galaxy, in accordance with Einstein's General Theory of Relativity.

paper was about measuring the size of molecules. For this, Einstein was awarded his doctorate from the Zurich Polytechnic. The third provided a theoretical explanation for Brownian motion – the movement of tiny particles suspended in liquid. Einstein used mathematical calculations to prove that the particles jiggle about as they do because the molecules in the liquid are in motion due to heat energy, and are colliding with the particles. This paper was important for providing further evidence of the existence of atoms.

Special Theory of Relativity

Einstein's fourth paper of 1905, 'On the Electrodynamics of Moving Bodies', was his most significant. This paper outlined his Special Theory of Relativity, which proposed that space and time are relative to the observer. In other words, the only reason why we all experience space and time in the same way is because we are all moving at the same speed, relative to each other. When observers move at very different speeds, strange things start to happen. For example, if someone on Earth was observing a passing spaceship that was travelling close to the speed of light, the spaceship would appear to grow shorter. If the Earthbound observer could measure the spaceship's mass, he would also find that it had become heavier. And if he could see a clock on the spaceship, he would notice that it was going slower than the clocks on Earth. Yet to the astronaut on board the spaceship, everything – the length and mass of the ship, and the progress of time – would appear normal.

The startling implication of this theory is that there is no such thing as absolute space and time – they depend on the position and speed of the object experiencing them. This had not been noticed before because at the slow speeds of our normal lives, the laws of classical physics – which state that space and time are absolute – appear correct. The only absolute, according to Einstein, is the speed of light, which is the same however and wherever it is measured. He also realized that nothing can go faster than light, because at that speed an object would have infinite mass, no length, and time would stand still.

$E = MC^2$

Soon after Einstein sent in his paper, he saw a further implication of his theory, and he immediately set to work on a fifth paper. Einstein had already stated that as a vehicle approaches the speed of light, its mass increases. To achieve this increase, energy is needed to push the vehicle faster. In other words, energy has been turned into mass. Einstein therefore concluded that mass is simply energy in a different form. From this he derived the famous equation $E = MC^2$ (energy equals mass times the speed of light squared). This was a completely new idea. Among other things, it explained how radiation worked. The equation could be used to show why a very large amount of energy could be emitted by a small piece of radioactive material (by converting mass to energy very efficiently). $E = MC^2$ also implied that there was a lot of potential energy contained within every atom.

Einstein's theories did not attract much attention at first. As a humble patent office clerk – even one with a doctorate – he lacked status in the scientific community. Also, his theories were so revolutionary and strange, and the equations used to reach them so complicated, that it is likely that many scientists did not fully understand them, or else rejected them as the work of a crank. Eventually Einstein received a letter from the renowned physicist Max Planck, asking some questions on relativity. This was followed in 1906 by a visit from one of Planck's assistants. Gradually, Einstein's ideas began to circulate. In 1907, as Einstein's reputation grew, he began looking for a university post so that he could continue his research. Two years later, he was offered the position of professor of theoretical physics at Zurich Polytechnic, and he was able to resign his job at the patent office. He spent a brief time at the German University in Prague, where he was awarded a full professorship, before returning to the Zurich Polytechnic in 1912. At the end of 1913, he was persuaded by Max Planck to join him as a professor at Berlin University. Here Einstein was free to continue his research, with only very light lecturing duties.

There is no such thing as absolute space and time – they depend on the position and speed of the observer.

General Theory of Relativity

This suited Einstein, who was engaged in an extension of his Special Theory of Relativity to include gravity. The special theory was so called because it only worked for objects that moved at a steady velocity, but not for objects that changed speed or direction because of gravity. Einstein eventually submitted his General Theory of Relativity in 1915 – a paper quite as remarkable as his special theory 10 years earlier. It stated that gravity is not a force – as physicists had believed since Newton –

but a distortion in space-time, created by the presence of mass. By 'space-time' Einstein meant that space and time, which we regard as separate things, are actually one four-dimensional continuum. Three of these dimensions are directions through space, and the fourth dimension is time. According to Einstein's General Theory, objects with mass create distortions, or curvatures, in space-time, and the larger the object, the greater the distortion. The planets orbit the Sun not because they are impelled to by a force, but because the Sun has curved space-time, causing the planets to follow an elliptical line through space.

Most people found Einstein's latest theory hard to fathom, and those few who could understand it rejected it as absurd. Einstein would need to provide some physical proof before the world would be ready to accept general relativity. The proof came a few years later. Einstein had said that everything would be affected by these distortions, even light. His theory would therefore be proved correct if he could show that the light from a star, viewed from Earth, bent as it passed around the Sun. The only time stars can be seen in daylight is during a solar eclipse. On 29 May 1919, the astronomer Sir Arthur Eddington went to Guinea in Africa to observe an eclipse. In November, the Royal Society of London announced that one of Eddington's photographs showed that a star whose light passed very close to the Sun appeared to shift position. The shift was almost exactly as Einstein had predicted.

The most famous scientist in the world

The announcement made headlines, and Einstein quickly became the world's most well-known scientist. He was deluged with letters, requests for articles and invitations to give lectures. Somewhat disturbed and embarrassed by all this acclaim, Einstein kept working. His next project was to try to find a link between electromagnetism and gravity. This was to be the first stage in a grand plan to discover a 'unified field theory' – a theory that could explain the laws that govern everything in the universe from subatomic particles to stars and planets. This quest, which was to dominate the rest of Einstein's life, was destined to fail. The new physics of quantum theory, which Einstein himself had helped to establish, showed that an uncertainty principle governed subatomic particles: mathematics can only predict where a particle probably is, not exactly where it is. Einstein recognized the validity of some aspects of quantum theory, but could never accept the uncertainty principle, or the use of probability as a means of solving problems in physics. As he said, 'God does not play dice with the universe.'

When Einstein published the first version of his unified field theory in 1929, it received a lot of attention from the world's press, but fellow scientists were critical. They claimed Einstein was heading in the wrong direction, and wished he would devote his efforts to helping them with quantum theory.

In the 1920s, Einstein became increasingly involved in political causes. A lifelong pacifist, he became an active campaigner for the cause. He travelled widely, and corresponded with such famous figures as the

The photoelectric effect

When light hits a piece of metal, electrons are dislodged from the atoms on the surface of the metal. Scientists knew about this photoelectric effect, but had no idea how it was caused. Einstein applied the recently developed quantum theory to the problem. Quantum theory showed that energy was emitted by radiating objects in discrete quantities, known as quanta. Einstein suggested that light behaved in the same way. A light beam, he said, was a stream of energy particles, which he called photons. Photons with sufficient energy could knock electrons from their atoms. Experiments performed in 1913 showed that Einstein was right.

A diagram showing a planet distorting the fabric of space-time, to illustrate Einstein's General Theory of Relativity. Space-time, represented here as a blue grid, is like a thin rubber sheet on which objects of varying weight produce smaller or larger dents.

psychiatrist Sigmund Freud and the Hindu poet and mystic, Rabindranath Tagore. In 1933 he accepted a post at a new Institute of Advanced Study at Princeton, New Jersey, in the USA. When he published a new version of his unified field theory in 1950, again it was criticized. By this time, his work was largely ignored by most theoretical physicists. Einstein died in April 1955, aged 76.

Einstein is remembered today as one of the greatest scientists who ever lived. His theories developed in the early years of the twentieth century, changed our understanding of the laws of the universe. They have been proved correct through observation and experiment time and time again. General relativity showed that the universe was expanding (even though at the time Einstein believed in a static universe) and this was proved by astronomer Edwin Hubble in 1929. His equation $E = MC^2$ had a practical application in the development of nuclear energy, and found a destructive one in atomic and hydrogen bombs. Today, physicists are still trying to link electromagnetism and gravity and complete Einstein's unified field theory.

Alfred Wegener

Alfred Wegener (1880–1930) first developed the theory that the continental land masses are constantly in motion, a theory now recognized as the most important and far-reaching development in the history of geology.

1880-1930

FOR THOUSANDS OF YEARS, PEOPLE BELIEVED THAT THE DIVISION OF LAND AND SEA was fixed and unchangeable. When Wegener suggested that this was not the case – that the hottest deserts had once been under the polar ice caps, and the countries differently distributed around the world – he struggled to get anyone to take him seriously.

Wegener was convinced that the moving continents were part of a mechanism that explained all the largescale activity of the Earth, including volcanoes, earthquakes, mountain-building and the movement of the magnetic poles. In this, he was finally proved right, but only long after his death. Trained in astronomy, and a meteorologist by profession, he drew from many disciplines to piece together evidence for his theory. But he was dismissed as an amateur, young and arrogant, propounding dangerous ideas.

Alfred Lothar Wegener was born on 1 November 1880 in Berlin, Germany. He was the youngest child of Dr Richard Wegener, an evangelical minister who ran an orphanage. From his early years, Wegener was fascinated by Greenland, later to become the arena of his most important work on meteorology and, finally, the scene of his death. He learned to ski and skate and trained vigorously during his youth, hoping one day to become a polar explorer.

Weather man

Wegener went to university in Berlin and achieved a PhD in astronomy in 1904, but quickly became interested in the new science of meteorology. Like his brother Kurt, Wegener took a job at the Royal Prussian Aeronautical Observatory near Berlin and soon made a name for himself. He used kites and balloons to study conditions in the upper atmosphere, and he and Kurt even broke a world record by staying aloft in a hot-air balloon for more than 52 hours.

In 1906 Wegener was delighted to be invited to join an expedition to the unmapped eastern coast of Greenland to study polar air flows. At last he could fulfil his dream to explore the polar wasteland. He became the first person to use kites and tethered balloons to study atmospheric conditions over the ice cap.

On his return to Germany he became a lecturer at the University of Marburg where staff and students alike were impressed by the clarity with which he could talk on even the most difficult subjects. He was known for quickly grasping new concepts,

The meeting of the North American tectonic plate and the European tectonic plate can be clearly seen near Þingvellir, Iceland, where ravines and cliffs mark the line of the Atlantic Fault. To the left of the picture is the eastern edge of the North American continental plate; to the right, the western edge of Europe.

integrating them seamlessly with what he knew already and seeing intuitively the answers to complicated problems. Yet despite being a rising star in the meteorological world, it was not in weather studies that he was to make his greatest impact.

Shifting continents

Wegener had begun thinking about the shapes of the continents early in his career. In December 1910 he wrote to the woman he would later marry: 'Doesn't the east coast of South America fit exactly against the west coast of Africa, as if they had once been joined? This is an idea I'll have to pursue.' Pursue it he did, and soon found evidence in the fossil record to encourage him.

In the autumn of 1911, Wegener came across a paper in the university library that listed fossils of plants and animals that are found on both sides of the Atlantic Ocean. He was intrigued, and set about finding more. At the time, scientists explained the similarities by suggesting that land bridges had once spanned the oceans but that these had sunk without trace as the Earth cooled and contracted. Wegener was not convinced by this argument. He suggested instead that there had once been a single, vast continent that had split up and the sections had drifted apart. He called this prehistoric continent Pangea, from the Greek 'all Earth'. Wegener found his theory immediately compelling. He wrote, 'A conviction of the fundamental soundness of the idea took root in my mind.'

He first presented his ideas at a meeting of the Geological Association in Frankfurt in January 1912, explaining to a startled audience how he saw the continents moving apart as the sea between them widened. He was quite aware that his theory required overturning all existing understanding of the history of the Earth.

The same year as he first presented his ideas on continental drift, Wegener set off to Greenland again. It was to be a dangerous trip, the four-man team only narrowly escaping death when a glacier they were scaling split beneath them. They became

The Origin of Continents and Oceans

Wegener proposed that the continental land masses, instead of being rooted deep in the Earth, are moving over it. Rock under the ocean is principally basalt, a denser rock than the granite which makes up the continents. He saw the land masses floating on this, rather like ice floats on the sea, though considerably slowed by the density of the rock they are forcing their way through: land masses ploughing through the oceanic crust like an icebreaker through a frozen sea. At the start of Earth's history, he thought, there was just a single large continent, which began to break up 200 million years ago, and the parts are still moving. Mountain ranges have been produced where one moving land mass crashes into another, pushing the rocks together and forcing them upwards in folds.

the first people to over-winter on the polar ice cap, and in the following spring made the longest journey over the ice sheet ever made, crossing 750 miles of snow and scaling icy peaks up to 10,000 feet high. The data that Wegener collected, and the work he did with it on his return to Germany, earned him respect as a world expert on polar meteorology and glaciology.

Back in Germany, Wegener married Else Köppen, daughter of the great meteorologist W. P. Köppen, and carried on working on his theory of continent drift. When Wegener first published his theory as *The Origin of Continents and Oceans* in 1915 it did not have the impact he had hoped for. The First World War had started, and though Wegener himself was quickly discharged from armed service after he was wounded, his theory went unnoticed outside Germany – the world was busy.

Controversy and conflict

It was not until 1922, when a third edition of his book was translated into several languages, including English, French and Spanish, that it attracted international attention. The response was not encouraging; the book was almost universally reviled, especially in the USA. The president of the American Philosophical Society summed up the general feeling, calling Wegener's theory 'Utter, damned rot!'

Wegener was invited to talk in New York about his theory, but met only hostility. Opponents were often downright rude, considering him an amateur in the field of geology, a man arrogantly straying outside his own area of expertise to tell them what to think.

It did not help Wegener's case that he had no plausible explanation for how the continents moved around over the surface of the Earth. During the 1920s he proposed that a force he called 'polflucht', produced by the Earth's spinning, caused land to pushed away from the poles. This, and some kind of tidal forces, might propel the continents on their course, he suggested. It was not convincing, even to Wegener. One opponent calculated that a tidal force strong enough to move continents would stop the Earth rotating within a year.

A few scientists supported Wegener's ideas. The Swiss geologist Émile Argand accepted the idea of colliding landmasses as a good explanation for the buckled, distorted strata he found in his studies of the Alps. South African geologist Alexander Du Toit was happy to believe that the similar fossils found in Africa and South America had been deposited when the land masses were adjacent. Helpfully, Alfred Holmes, a professor at Edinburgh University, suggested that convection currents deep within the Earth might move the continents, a theory Wegener included in his 1929 edition of the book and which is now generally accepted. For the majority, though, it was an absurd idea supported by little real evidence and with nothing to recommend it over existing theories.

For such a radical theory to be generally accepted, Wegener knew that he would need a lot of supporting evidence. He looked for this in different disciplines, studying geological features on both sides of the Atlantic as well as the fossil record. Mapping bands of mountains, and of deposits of coal and minerals, shows continuous strips that run across from one continent to another – from Africa to South America, and from Antarctica through India to Africa, for instance.

Wegener's most compelling evidence came from paleoclimatology – the study of climate patterns millions of years ago. With the help of Vladimir Köppen, he plotted ancient jungles, ice sheets and deserts on his map of Pangea. It all made sense.

The permo-carboniferous ice age, which occurred 280 million years ago, had previously appeared to show the ice sheet scattered randomly around the world, some of it in the hottest deserts. On Wegener's map, it centred in one place near the south pole, where Africa, Antarctica, Australia and India once met.

A tragic, heroic end

No German university would appoint Wegener as a professor because of the controversy surrounding his ideas. Luckily, a professorship in meteorology and geophysics was created for him in 1924 in the small university of Graz, Austria.

From Graz he continued to work on polar weather, and returned to Greenland in 1930. The expedition went badly from the start. When part of his expedition was stranded 250 miles from the coast, Wegener, as leader, had to launch a rescue mission. In the atrocious conditions, most of Wegener's rescue team turned back. It took Wegener and two companions forty days to cross the ice to the stranded camp, in temperatures as low as –58 degrees.

The day after celebrating his fiftieth birthday in 1930, Wegener and Villumsen, his Greenlander helper, set out on the return journey. They never arrived. In the spring, Wegener's body was found stitched into his sleeping bag and marked by two skis upright in the snow. Villumsen had buried him and gone on, but he had disappeared without trace in the icy wilderness. Wegener's team erected an ice mausoleum and later a 20 foot iron cross where his body lay. Both have since vanished under the snow and ice.

Wegener's theory languished after his death. With no champion, only a few enthusiasts kept it alive. Yet in the 1950s new scientific methods made it possible to look at the activity of the Earth's crust in new ways. Examination of the sea floor (oceanography) and studies of how the magnetic polarity of the Earth has shifted over millions of years (paleomagnetism) began to throw up evidence in support of Wegener's theory of moving land masses.

Plate tectonics

Modern plate tectonics theory explains the movement of the land masses suggested by Wegener. The top part of the Earth – the crust and top level of the mantle – is divided into seven large and several more small plates. These float on top of the rest of the mantle, which is formed of thick, sticky, liquid rock at a high temperature (magma).

The plates slowly move around the Earth, and their movement accounts for the continental drift Wegener described – though he was wrong to assume only the land plates moved, or that they moved through the oceanic crust. We can now trace the past movement of the plates over millions of years, and have established that the Atlantic Ocean is indeed still growing, though North America is moving away from Europe at only 2.5 cm a year – one hundredth of the rate Wegener suggested.

There is still no consensus as to why the plates move over the asthenosphere, though the favourite theory, proposed by Holmes, is that they are carried by convection currents in the magma beneath. As Wegener said, 'The Newton of drift theory has not yet appeared. His absence need cause no anxiety; the theory is still young'.

Niels Bohr

Niels Bohr (1885–1962) was one of the greatest physical chemists of the twentieth century. He proposed the 'solar system' model of atomic structure, in which electrons orbit the central nucleus – a model which still underpins our understanding of matter.

1885-1962

Bohr was an inspired thinker with immense powers of concentration. He was dedicated to his work and was known for his staying power, but was also a generous and softly spoken man. His promise was evident early in his life; a fellow student wrote of him in 1904: 'It is very interesting to know a genius and I do, I am even together with him every day. I am talking about Niels Bohr … and besides he is the best, most modest human being that can be imagined.' His humanitarian spirit came to the fore in later life when he, like many other scientists of his day, campaigned against indiscriminate nuclear arms development.

Bohr won a Nobel Prize for physics for his groundbreaking work on the structure of the atom. The element bohrium is named after him, and the Institute for Theoretical Physics in Copenhagen, which he headed in his lifetime, was renamed in his honour.

Early life

Niels Bohr was born on 7 October 1885 in a stately mansion owned by his maternal grandmother, a woman from a wealthy and influential Jewish banking family. His father, Christian Bohr, was appointed professor of physiology at the University of Copenhagen. Niels, his brother and sister grew up in a stimulating atmosphere in which the pursuit of knowledge was respected and encouraged, surrounded by intellectual discussion and culture. Bohr later said that philosophical discussions amongst his father's friends at the house inspired him to look for unifying principles in human knowledge – a quest which he surely fulfilled in applying quantum theory to chemistry.

Bohr was not academically exceptional at school, usually coming third or fourth in a class of twenty. He was a skillful footballer, but in this he was outshone by his brother Harald, who played for Denmark in the 1908 Olympics, winning a silver medal. Despite keen competition on the football field, Niels and Harald were best friends, and remained inseparable all their lives. Some of Bohr's discoveries were first related to his brother, in the letters they exchanged frequently.

A promising start

Bohr studied at the University of Copenhagen, but because there was no physics laboratory in the university, he was only able to carry out experimental work by using his father's physiology laboratory. Even so, in 1906 he won the Gold Medal from the Royal Danish Academy of the Sciences for his measurement of the surface tension of water.

Bohr completed his PhD in 1911 and later in the same year went to England, intending to work with J.J. Thomson at the University of Cambridge. However, the two did not get on well and Bohr looked for an escape route.

He was fortunate to find one quickly. Ernest Rutherford had just published his discovery that most of the mass of an atom is in its nucleus (the centre). He was in Cambridge to give a talk on his work in October 1911, and Bohr heard him and was greatly impressed. When Bohr went to Manchester a month later to visit a friend of his father, Rutherford was invited to dinner. The meeting was a success, and in March of the following year Bohr joined Rutherford's team in Manchester working on the structure of the atom. He adopted Rutherford as something of a role model, both professionally and personally. The two became firm friends, though they had very different characters, and after Bohr left Manchester they wrote to each other frequently until Rutherford's death in 1937.

A new atom

While at Manchester, Bohr worked with quantum theory developed by Einstein and Planck to explore his own theories about atomic structure and fix the faults he could see in Rutherford's model. Although Rutherford's model was a brilliant innovation and represented a huge leap forward, it didn't quite work. In Rutherford's atom, the electrons would slowly spiral into the middle, or could be knocked out of position by a nearby positive particle.

Bohr left Manchester after 6 months and returned to Copenhagen where he married his fiancée Margrette Nørlind in the summer of 1912. They were to have

The quantum atom

The key difference in Bohr's model of atomic structure was that the electrons occupied distinct orbits, or shells, rather than whirling arbitrarily around the nucleus in a cloud.

An electron can jump between orbits but is never in between two orbits. Jumping to an orbit further from or closer to the nucleus is associated with absorbing or giving out energy.

The innermost orbit contains up to 2 electrons. The next may contain up to eight electrons. If an inner orbit is not full, an electron from an outer orbit can jump into it. Energy is released as light (a photon) when this happens. The energy released is a fixed amount, a quantum.

Hydrogen emission spectra – lines showing the light given out by hydrogen when bombarded with alpha-particles – provided evidence for Bohr's model. The emission spectra show light is emitted in regular patterns as the hydrogen molecules' electrons move between orbits.

A diagram of Bohr's model of the atom, with electron shells.

six sons, two of whom died young. The fourth, Aage, would eventually follow his father into physics and win his own Nobel Prize in 1975.

Back in Copenhagen, Bohr carried on working on his theory of the atom, publishing his theories in three papers in England in 1913. It was for this explanation of atomic structure that he was awarded the Nobel Prize for Physics in 1922. His work became the foundation of quantum mechanics, which developed during the 1920s, largely centred around his institute in Denmark.

In 1914, Bohr was to take up a new a professorship in theoretical physics in Copenhagen, but the start of the First World War delayed the creation of the post until 1916. In the meantime, Rutherford offered him a readership for 2 years at Manchester. He and his wife endured a dangerous journey by sea in the middle of the war to take up the position. The readership gave him the opportunity to continue with his research without having to devote time to teaching in the university.

On returning to Denmark and his new professorship, Bohr was elected to the Royal Academy of Sciences. In 1921 he became president of the newly established Institute of Theoretical Physics (sponsored by the Carlsberg brewery), which he had petitioned to open. He held this post until his death, and the institute was later renamed in his honour. His son Aage succeeded him as director in 1963.

Unravelling the elements

Besides his work on quantum theory, Bohr pursued the implications of his model of atomic structure for the periodic table of elements. He showed that the characteristics of an element could be accounted for and even predicted by the configuration of electrons in its atoms, and so by its position in the periodic table.

Under Bohr's guidance, the institute in Copenhagen attracted some of the leading physicists from abroad and became a world centre for work on atomic physics and quantum theory. Bohr himself travelled and lectured in Europe, the United States and Canada.

All change

In the mid-1920s, Bohr stressed that the new models of the atom on which he was working were theoretical – he foresaw a new shift and synthesis of ideas coming together. It arrived the same decade with the emerging field of quantum mechanics, which was grounded in Bohr's quantum model of the atom. In 1927, Heisenberg published his uncertainty principle, which says that it is impossible to measure the position and energy of a particle, since the act of measuring will affect the particle and so alter its state.

Atomic structure and the chemistry of the elements

Each of the elements has an atomic number, starting with hydrogen, with an atomic number of one. The atomic number corresponds to the number of protons in the element's atoms.

Bohr had already shown that electrons inhabit fixed orbits around the nucleus of the atom. Atoms strive to have a full outer shell (allowed orbit), which gives a stable structure. They may share, give away or receive extra electrons to achieve stability.

The way that atoms will form bonds with others, and the ease with which they will do it, is determined by the configuration of electrons. As elements are ordered in the periodic table by atomic number, it can be seen that their position in the table can be used to predict how they will react.

In September 1927, Bohr took account of Heisenberg's principle in explaining the concept of complementarity (see box).

Albert Einstein was doubtful about Bohr's new interpretation of quantum theory, though in the end Bohr's version prevailed. The two men debated the issue over many years, and although Einstein never agreed with him, Bohr acknowledged the huge value of their discussions in refining his ideas. In 1927 he wrote: 'Anyone who is not shocked by quantum theory does not understand it.'

Complementarity

Bohr's theory of complementarity states that electrons may be both a wave and a particle, but that we can only experience them as one or the other at any given time. He showed that contradictory characteristics of an electron can be proved in separate experiments and none of the results can be accepted singly – we need to hold all the possibilities in mind at once. This requires a slight adjustment to Bohr's original model of atomic structure, in that it means we can no longer say that an electron occupies a particular orbit, but can only give the probability that it is there.

A difficult war

During the 1930s Bohr became interested in nuclear fission and the possibility of gaining energy from it. Nuclear fission involves the splitting of an atomic nucleus, causing a release of energy. Work on nuclear fission rapidly became part of the race to develop an atomic bomb as the Second World War unfolded. Lise Meitner, who had escaped from Nazi-occupied Austria, brought news that the Germans were researching nuclear fission. And on a visit to Bohr, Heisenberg revealed that Germany was working on an atomic bomb – indeed, Heisenberg was in charge of the project. He later claimed that he and Bohr came to an understanding that Heisenberg would undermine the project if it looked as if it would succeed, but Bohr denied such an agreement was ever made.

When Hitler began persecuting Jews in Germany, Bohr offered a safe haven at the Institute in Copenhagen for many Jewish scientists, and after the outbreak of war he even donated his gold Nobel medal to the Finnish war effort. When the Germans invaded Denmark in 1940, Bohr's Jewish descent made life difficult for him, especially as he made no secret of his anti-Nazi feelings. In a daring escape, he and his family fled to Sweden in a fishing boat provided by the Resistance movement. From there they went to England, hiding in the empty bomb rack of a British Mosquito plane sent to pick them up.

'Anyone who is not shocked by quantum theory does not understand it.'

Safely in England, Bohr joined the war effort to develop the atom bomb ahead of the Germans. He and his son Aage later moved to Los Alamos in the United States with the rest of the British research team to join the Manhattan Project. But Bohr was not happy about the development of nuclear weapons, and in 1944 tried to persuade both Roosevelt and Churchill that international cooperation would be a better path forwards in the development of nuclear fission. Churchill was annoyed that Bohr thought their knowledge should be shared with the Russians and that he favoured post-war arms control. In 1950, Bohr wrote to the United Nations to put his case against unilateral development of nuclear arms.

In 1955 Bohr organised the Atoms for Peace Conference in Geneva. He was also a leading figure in the foundation of CERN, the Centre for Nuclear and Particle Physics Research in Switzerland, founded in 1954. He died in Copenhagen after a stroke on 18 November 1962.

Edwin Hubble

The American astronomer Edwin Powell Hubble (1889–1953) was the man who almost single-handedly changed the view of the universe as being of limited size, and laid the foundations for our modern understanding of the cosmos.

1889–1953

At the start of the twentieth century, astronomers believed that our galaxy, the Milky Way, was the entire universe, measuring just a few thousand light years across. In the 1920s, Erwin Hubble revealed that the Milky Way was only one of billions of galaxies in a universe of vast dimensions. He also discovered that the universe was expanding, thus providing the first evidence for the 'big bang' theory.

Early life

Hubble was born on 29 November 1889 in Marshfield, Missouri, USA. When he was 9, his family moved to Wheaton, Illinois, a suburb of Chicago. He grew up tall and strong, and became an outstanding athlete, regularly coming first in school sports events. In 1906, he even broke the Illinois state record for the high jump.

Hubble also showed a strong aptitude for science, gaining a place at the University of Chicago to study mathematics and astronomy. Here he was particularly inspired by the renowned astronomer George Ellery Hale. When not studying, Hubble kept up his sporting pursuits, playing for the university basketball team. He was also a talented boxer – so much so that boxing promoters tried to persuade him to turn professional. Fortunately for the future of astronomy, Hubble refused the offer.

This combination of academic and sporting achievement earned Hubble a Rhodes scholarship to Oxford University in 1910. Despite his love of science, Hubble made a promise to his dying father that he would pursue a legal career, and he studied law at Oxford. On his return to America, aged 23, Hubble contemplated a career as a lawyer, but ended up finding work as a basketball coach and high-school teacher. His time in England had left its mark on Hubble, who took to dressing like an Oxford scholar, smoking a pipe and speaking with a British accent (or his own version of it). Although popular with his students, who were charmed by his eccentricities, Hubble longed to return to science.

He enrolled as a graduate student at Yerkes Observatory in Wisconsin, where he began studying the faint, cloudy formations known as nebulae that would one day make him famous. In 1917 he received his doctorate in astronomy from the University of Chicago. His evident skill as an astronomer won him an offer from the prestigious Mount Wilson Observatory near Pasadena, California.

Cepheid variables

Hubble arrived at Mount Wilson in 1919, aged 30. At this time, astronomers believed there was just one galaxy in the universe – the Milky Way (the word *galaxy* actually comes from the Latin for 'milky vault'). However, some recent progress had at least been made in terms of understanding the dimensions of the universe. A woman named Henrietta Swan Leavitt, who worked at Harvard College Observatory, had discovered a type of star known as a Cepheid variable (after the constellation Cepheus where the first one was found). These stars brightened and dimmed in a regular rhythm. (Cepheid variables are now known to be 'red giants' – very old stars.) Leavitt realized that a relationship existed between their brightness (or closeness to us) and the speed of their rhythm. By comparing the relative brightness of Cepheid variables at different points in the sky, she could work out where they were in relation to us, and to each other. For the first time it had become possible to measure distances between different parts of the universe.

Before Hubble joined the team at Mount Wilson, another astronomer there named Harlow Shapley had already startled the world with his conclusions about the size of the Milky Way. Using Cepheid variables to measure the distances, Shapley had judged it to be 300,000 light-years across – ten times bigger than was previously thought. But Shapley, like most other astronomers of the time, believed the Milky Way was all there was. The strange clouds called nebulae were, he said, merely puffs of gas.

Discovering new galaxies

Hubble was lucky enough to arrive at Mount Wilson just after the observatory had built the 2.54 m Hooker Telescope, the most powerful on Earth. He was therefore able to observe the heavens in far greater detail than his predecessors. After a few years of patient observation, he made an extraordinary discovery. In 1923, he spotted a Cepheid variable in one of the so-called puffs of gas known as the Andromeda Nebula. Using Leavitt's technique, he was able to prove that Andromeda was nearly a million light-years away – far beyond the outer limits of the Milky Way, and was clearly a galaxy in its own right.

The light from stars Hubble was looking at displayed signs of red shift: they were moving away from the Earth.

Hubble went on to discover Cepheids in other nebula, and conclusively proved that galaxies existed beyond our own. He described his findings in a 1924 paper entitled 'Cepheids in Spiral Nebulae'. Almost overnight he became the most famous astronomer in the world. People suddenly had to get used to the fact that the universe was vastly bigger than anyone had previously imagined. Shapley, for one, was quite shaken by the news. He wrote to Hubble following his discovery: 'I do not know whether I am sorry or glad to see this break in the nebular problem. Perhaps both.'

In 1926, Hubble began to develop a classification system for the galaxies he had discovered, sorting them by content, distance, shape and brightness. In the course of his studies, he noticed an odd fact: they appeared to be moving away from the earth. Hubble knew this because the light from stars he was looking at displayed signs of something called red shift, in which the light waves from an object moving away at great speed from a stationary observer become stretched out as a result, and the light shifts towards the red end of the spectrum. Similarly, approaching light shifts to blue.

Galaxies expanding from the big bang. The big bang is a theory of the formation of the universe, in which the universe expanded from an infinitely dense point that existed 10–20 billion years ago.

Hubble was not in fact the first to notice this red shift in the emission of light from distant stars. In 1914, an American astronomer named Vesto Slipher had noticed the same thing, but his findings drew little or no attention. Hubble, making his observations over a decade later, had the advantage of a larger telescope, as well as the knowledge that the universe contains more than one galaxy.

An expanding universe

With his assistant Milton Humason, Hubble began measuring the distances to the receding galaxies, and by 1927 he was able to formulate what came to be known as Hubble's Law: the greater the distance of a galaxy, the faster it recedes. The inescapable conclusion of all this was that the universe, which had always been considered static, was actually expanding.

Two years later, Hubble calculated the rate of expansion, known as Hubble's constant (*H*). This enabled astronomers to work out the speed (v) at which any given galaxy was receding (v = *H* x distance). Hubble actually overestimated the size of his constant, basing it on the assumption that the Milky Way was the largest of all galaxies and that the universe was far younger than it actually is. However, his formula remains valid, and once subsequent astronomers had revised the constant, they were able to use it to calculate the size and age of the universe. Its radius has been estimated to be a maximum of 18 billion light-years, and it is believed to be between 10 and 20 billion years old.

Hubble's dramatic findings attracted the attention of the famous physicist Albert Einstein. In 1915, Einstein had put forward his General Theory of Relativity, which had suggested that, owing to the effects of gravity, the universe was either expanding or contracting. Yet the consensus among astronomers at that time was that the universe was static, and Einstein did not know enough about astronomy to disagree with them. So he introduced an anti-gravity force into his equations, which he called the cosmological constant. Hubble's discoveries had proved that Einstein's instincts had been right after all. Einstein later described the introduction of the cosmological constant as the 'biggest blunder of my life'. He even visited Hubble at Mount Wilson in 1931 to thank the astronomer for revealing his error.

Hubble's status as a scientific superstar was confirmed in 1936 with his book *The Realm of the Nebulae* in which he described how he made his discoveries. The Mount Wilson Observatory became a popular tourist destination, and Hubble was embraced by the elite of California society.

When the Second World War broke out in December 1941, Hubble was determined to fight on the front line. However, he was persuaded that he would be more useful to his country working behind the scenes as a scientist, and he was made head of ballistics at a research centre in Maryland.

Before and after the war, Hubble played a central role in the design and construction of the Hale Telescope on Palomar Mountain, California. Completed in 1948, the 5.08 m Hale was four times as powerful as the Hooker, and would remain the largest telescope on Earth for the next 40 years. Hubble had the honour of being the first to use it. Asked by a reporter what he expected to find, Hubble replied: 'We hope to find something we hadn't expected.'

The Hubble Space Telescope

Perhaps more than anyone else, Edwin Hubble expanded our understanding of the universe. It is therefore fitting that today he is remembered by the Hubble Space Telescope (HST), an orbiting observatory that has shown us some of the most stunning views of the cosmos ever observed. The Earth's atmosphere alters light rays from outer space, giving ground-based telescopes, however powerful they are, a distorted view of the heavens. The HST, being above the Earth's atmosphere, receives images with much greater clarity and detail.

Construction began on the HST in 1977, and it was launched into space by the space shuttle *Discovery* on 25 April 1990. Its instruments can detect not only visible light, but also ultraviolet and infrared light. Its camera is able to achieve a resolution ten times greater than even the largest Earth-based telescope. As a result, today's astronomers can observe distant celestial objects with a clarity that Hubble and his contemporaries could only have dreamt of.

Later life

Hubble received many awards for his achievements. In 1946 he was awarded the Medal of Merit, and in 1948 he was elected Honorary Fellow of Queen's College, Oxford. However, the one honour that always eluded him was the Nobel Prize. Unfortunately for Hubble there was no prize for astronomy, and by the time the Nobel committee decided that astronomy was a branch of physics, it was too late: Hubble had died.

Hubble continued to work at both the Mount Wilson and Mount Palomar observatories until his death from a cerebral thrombosis on 28 September 1953. His legacy to astronomy is immense. He transformed our view of the cosmos and our place in it. His discovery that the universe is expanding led scientists to develop the 'big bang' model, which remains the most compelling theory about the origin of the universe. According to this theory, the universe began some 10–20 billion years ago, exploding outwards from a tiny point of almost unimaginable heat and density, and has been moving apart ever since.

Hubble was a brilliant observational astronomer who tended to leave the theoretical implications of what he observed to others. He showed what was possible to achieve simply by looking through the viewfinder of a telescope. As Hubble himself said: 'Equipped with his five senses, man explores the universe around him and calls the adventure Science.'

Werner Heisenberg

Werner Heisenberg (1901–76) found a new way of expressing the paradoxical nature of the subatomic world, using mathematics. In doing so, he laid the foundations for a new branch of physics, known as quantum mechanics. He received the Nobel Prize for his work in 1932.

1901–76

In his 1963 book *Physics and Philosophy*, Werner Heisenberg noted: 'We wish to speak in some way about the structure of atoms.... But we cannot speak about atoms in ordinary language.' In this statement, the German physicist and philosopher articulated the problem facing scientists in the early twentieth century: that electrons and other subatomic particles did not possess a physical form that could be visualized or described in words – sometimes they behaved like particles and at other times like a wave. His solution was matrix mechanics, developed in 1926, from which he derived his famous 'Uncertainty Principle' (1927).

Werner Karl Heisenberg was born on 5 December 1901 in Würzberg, Germany. He was the younger of two sons born to Dr August Heisenberg, a scholar of classical languages, and his wife Annie Wecklein. In 1910, the family moved to Munich, where Heisenberg attended the Maximilian Gymnasium. He was an excellent all-round student, scoring his highest marks in mathematics, physics and religion.

In 1920, he enrolled at the University of Munich to study physics under the world-renowned physicist Arnold Sommerfeld. Here, Heisenberg met his lifelong friend and colleague, Wolfgang Pauli. In 1923, after completing their doctoral dissertations, Heisenberg and Pauli went to the University of Göttingen, where they studied under the quantum theorist Max Born.

> *We wish to speak in some way about the structure of atoms.... But we cannot speak about atoms in ordinary language.*

Bohr's atomic model

Then, in the autumn of 1924, Heisenberg went to the University of Copenhagen to study under Niels Bohr, the Danish physicist famous for his work on the atom. Since 1912, Bohr had been at the forefront of developments in quantum theory. This theory describes the behaviour of subatomic particles based on the assumption that they behave in both a wave or particle-like fashion.

Heisenberg was very interested in Bohr's model of the atom, which for the first time attempted to incorporate quantum theory. Earlier 'solar system'-type models of the atom had electrons simply orbiting the central nucleus like planets going around a sun. Bohr agreed that electrons orbit the nucleus, but he pointed out that, unlike solar systems, the energies of the electrons can only occur in fixed amounts, or quanta. These quanta corresponded to certain fixed orbits. An electron could jump from one fixed orbit to another by emitting or absorbing energy exactly equal to the difference in energy between the orbits. Visualizing the atom in this way made sense of what people saw when they looked at the spectral lines of a hydrogen atom. Spectral lines are obtained by directing the electromagnetic radiation (caused by the vibrations of electrons) of an element through a prism. The prism breaks it up into spectral lines, which show the intensities and frequencies of the radiation – and therefore the energy emissions and absorptions of the electrons.

The Bohr model had its flaws, however. Although it contained elements of quantum theory, it still ignored the wave character of the electron. Also, Bohr's model only worked for single-electron hydrogen atoms. Beginning in April 1925, Heisenberg decided to try and develop a new model of the atom, more fundamentally based on quantum theory, that worked for all atoms. He believed the approach of trying to visualize a physical model of the atom was destined to fail because of the paradoxical wave-particle nature of electrons. Since the orbits of electrons could not be observed, he decided to ignore them and focus instead on what could be observed and measured, namely, the energy they emitted and absorbed as shown in the spectral lines. He tried to devise a mathematical way of representing the orbits of electrons, and to use this as a way of predicting the atomic features shown up in the spectral lines.

Matrix mechanics

By July, Heisenberg had come up with something that seemed to work, but the mathematics was so abstract and strange that he was not sure it even made sense. It involved arrays of numbers, or matrices. (A matrix is a set of mathematical equations whose rows and columns can be combined with other matrices to solve problems.) He showed it to Max Born, who recognized it as conforming to a form of mathematics called matrix algebra. With the help of Born and another physicist, Pascual Jordan, Heisenberg fine-tuned his theory, which he named matrix mechanics. Further experimentation showed that matrix mechanics could account for many of the properties of atoms, including those with more than one electron.

While physicists were impressed with the effectiveness of matrix mechanics as a means of predicting subatomic behaviour, they were rather repelled by the obscure nature of the mathematics involved, and by the fact that it made it no easier to visualize what an atom might actually look like. In early 1926, an Austrian physicist called Erwin Schrödinger came up with an alternative theory called wave mechanics. In his theory, Schrödinger

The second UN International Atomic Energy Conference at Geneva, September 1958. Werner Heisenberg is second from right.

stated that the quantum energies of electrons did not correspond to fixed orbits, as Bohr had stated, but to the vibration frequency of the 'electron-wave' around the nucleus. Just as a piano string has a fixed tone, so an electron-wave has a fixed quantum of energy. Wave mechanics used much simpler mathematics than matrix mechanics and was also easier to visualize. In May 1926, Schrödinger showed that in mathematical terms, both theories were the same, but that – in his view – wave mechanics was more elegant and accessible. The rival theories together formed the basis of what became known as quantum mechanics.

In October 1926, while Heisenberg began a new job as a lecturer at the University of Copenhagen, Schrödinger arrived in the city to debate the alternative theories with Bohr. The debates were passionate, but inconclusive. They only made clear the unsatisfactory nature of both theories. Heisenberg realized that his mathematical formulation was not sufficient. He would have to come up with a way of interpreting his matrices that made sense in physical terms.

The Uncertainty Principle

In the meantime, Pascual Jordan, together with the English physicist Paul Dirac, created a new set of equations based on the rival theories, which they called 'transformation theory'. While studying the Dirac–Jordan equations, Heisenberg noticed a problem. Whenever one tried to measure both the position and velocity (speed and direction) of a particle at the same time, the results were imprecise or uncertain. Heisenberg believed that this uncertainty was not the fault of the equations, but was part of the very nature of the subatomic world. He wrote a fourteen-page letter to Pauli in February 1927, explaining his new theory, and this formed the basis of a published paper. He called it the 'Uncertainty Principle'.

The theory states that one can know the position of a subatomic particle at a particular instant or one can know its velocity, but one cannot know both at the same time. The reason for this is that the very act of measuring the velocity of a subatomic particle will change it, making the simultaneous measurement of its position invalid. The principle also works in the visible world, but we don't notice it because the uncertainty element is extremely tiny. It is, for example, extremely easy to work out both the position and velocity of a car at any given moment. However, in the tiny subatomic world, the uncertainty element becomes extremely significant.

One can know the position of an subatomic particle or one can know its velocity, but never at the same time.

This is not something that can be solved with more precise measuring techniques or instruments: it is to do with the fundamental relationship between particles and waves at the subatomic level. Every particle has an associated wave. The position of a particle can be precisely located where the wave's undulations are most intense. But where the wave's undulations are most intense, the wavelength is also at its most ill-defined, and the velocity of the associated particle is impossible to determine. Similarly, a particle with a well-defined wavelength has a precise velocity but a very ill-defined position.

An unobserved object is a mixture of both particle and wave. If an experimenter chooses to measure the object's velocity, the object will transform itself into a wave. And if the experimenter chooses to measure its position, it will become a particle. The velocity and position of the object were indeterminate – they did not exist –

before they were observed. Therefore, by choosing to observe either one thing or the other, the observer is actually affecting the form the object takes.

The practical implication of this is that one can never predict where an electron will be at any given moment; one can only predict the probability of its being there. In another sense one can say that an electron does not actually exist – or at least it exists in an undefined state – until it is observed.

Bohr was away on a skiing trip when Heisenberg began writing up his paper on the Uncertainty Principle. He showed Bohr his first draft on his return, and the Danish physicist was so impressed he immediately sent a copy to the famous Swiss–German physicist Albert Einstein for his comments. Einstein did not like the theory's reliance on probabilities. He was even more opposed to the idea that the observer could influence what he observed. For Einstein, nature existed independently of the experimenter. Despite the great physicist's objections, the Uncertainty Principle quickly gained support. Bohr made use of it in coming up with his Complementarity Principle in 1928, which states that a complete understanding of a subatomic object requires a description of both its wave and particle properties. Uncertainty and Complementarity together became known as the 'Copenhagen interpretation' of quantum mechanics.

Later years

Along with Bohr, Max Born and other supporters of the Copenhagen interpretation, Heisenberg began gathering support for the doctrine by presenting it at lectures around the world. By the early 1930s, it had become widely accepted by physicists everywhere. Students from places as far away as America, India and Japan flocked to hear Heisenberg's lectures. However, notable dissenters to the doctrine remained, including Einstein, Schrödinger and the French physicist Louis de Broglie. In 1932, Heisenberg was awarded the Nobel Prize for physics for his contributions to quantum mechanics.

The Nazi Party came to power in Germany in January 1933. Heisenberg remained in Germany throughout the period of the Third Reich, despite his dislike of the Nazis. He was a patriot who believed it was important to remain in Germany to help preserve the status of German science. Throughout the 1930s, he continued to work, using quantum mechanics to investigate solid-state crystals, the structure of molecules and the scattering of radiation by nuclei.

In his later years, following the Second World War, Heisenberg played a leading role in establishing the European Council for Nuclear Research (CERN) in Switzerland in 1952, and became very involved in research into high energy physics. In the 1960s, he turned most of his energies to writing and lecturing. He died on 1 February 1976.

Heisenberg and nuclear fission

In 1938, German scientists Otto Hahn, Lise Meitner and Otto Frisch discovered nuclear fission – the splitting of the nucleus of a uranium atom and consequent release of huge amounts of energy. The following year, the Second World War broke out, and Heisenberg and his colleagues were asked to look at military applications for nuclear fission. Heisenberg later defended his involvement in the project, saying he did it out of self-protection (he had been threatened by the Nazis before the war), and would have sabotaged the project had they succeeded in actually creating an atomic bomb. As it happened, they never came close to building such a device. After the war, the occupying authorities banned nuclear fission research in Germany, and Heisenberg led a campaign to lift the ban. He saw the peaceful development of nuclear energy as a necessary part of the revival of the German economy, while remaining fundamentally opposed to nuclear weapons. The ban was eventually lifted in 1955.

Linus Pauling

Linus Pauling (1901–94) is considered to be the most influential chemist since Lavoisier and the founding father of molecular biology. His work on chemical bonds – the forces that hold atoms together in molecules and crystals – formed the basis of our understanding of molecular structures.

1901-94

ALTHOUGH LINUS PAULING PLAYED A KEY ROLE in unravelling the chemicals that make life possible, this is only half of the story. He was also a great humanitarian and a devoted worker for world peace and civil liberties. He won the Nobel Prize for chemistry in 1954 for his work on the chemical bond and in 1962 he won the Nobel Peace Prize, making him the first person to win two unshared Nobel Prizes.

Pauling, unlike most other great scientists, is not famous for one or two key discoveries. Instead, his influence extended far and wide, and we can think of him, perhaps, as stitching together all the major work of the twentieth century in chemistry and biochemistry.

An early start in chemistry

Linus Carl Pauling's childhood was unsettled and marked by poverty and tragedy. He was born in Portland, Oregon, on 28 February 1901. His father was a failed pharmacist of German descent who died when Pauling was only 9 years old. Thereafter his mother brought up her 3 children on her own. Pauling had to work from the age of 13 to help support the family.

Despite this hardship, Pauling was a bright and curious child who read voraciously – his father once wrote to a local paper asking for suggestions for books his son could read. He showed an early interest in science, particularly chemistry, and loved experimenting in the small laboratory that a friend, Lloyd Jefress, kept in his bedroom. But Pauling failed to gain his high school diploma because he refused to take an American history course in the way the school required. The diploma was withheld until after he had won two Nobel Prizes, when the school finally relented.

After high school, Pauling went to Oregon State Agricultural College, Corvallis, in 1917 to study chemical engineering. His mother was living in poverty, and Pauling had to work full time while studying. From 1919 to 1920 he taught the course in analysis that he had just finished taking, earning the epithet 'the boy professor'. It was all that saved him from having to leave and return to Portland to support his mother. After finishing his degree, he moved to the California Institute of Technology ('Caltech') in Pasadena where he taught from 1922 to 1925 and gained his PhD in chemistry in 1925.

In 1923 he married Ava Helen Miller, a brilliant young woman whom he met and fell deeply in love with while teaching a course in chemistry for home economics. Ava was a passionate campaigner for women's rights and later worked with Pauling on his campaigns against nuclear arms. Together they had three sons and a daughter.

Atoms and bonds

Already, in 1919, Pauling was interested in the way atoms form bonds with one another. It was to become the main thrust of his career as a chemist. He was inspired by the work of Irving Langmuir and Gilbert Lewis, who had suggested that pairs of electrons are shared between atoms, holding the atoms together. Pauling began work on crystalline structures at Caltech in 1922, hoping to find out why atoms in metals arrange themselves into regular patterns.

Pauling had learned about X-ray diffraction as a student and used it to determine the crystalline structure of molybdenum in 1922. When X-rays are directed at a crystal, some are knocked off course by striking atoms, while others pass straight through if there are no atoms in their path. The result is a diffraction pattern – a pattern of dark and light lines that reveal the positions of the atoms in the crystal.

Linus Pauling, shown in a photograph from 28 April 1962, holding up a sign as he joined pickets in front of the White House during a mass protest against resumption of US atmospheric nuclear testing.

Pauling studied many more crystals, and in 1928 published his findings as a set of rules for working out probable crystalline structures from X-ray diffraction patterns.

Inspiration from Europe

After completing his PhD in 1925, Pauling travelled to Europe for 2 years on a Guggenheim Fellowship, where he worked with some of the most influential figures of the day, including Niels Bohr, Edwin Schrödinger and William Bragg. Most inspiring for him, though, was his observation of the work of Fritz London and Walter Heitler on the quantum mechanics of the hydrogen atom.

Returning to the United States, Pauling became one of the first people to apply the new theories of quantum mechanics to molecular structure with a view to understanding chemistry. He used it in a comprehensive study and explanation of bonds between atoms.

'The nature of the chemical bond'

It was already known that atoms could combine with others forming bonds that were either ionic or covalent. Pauling was to do away with this neat but over-simple classification.

He began publishing his work on chemical bonds in 1931 after a second visit to Europe in which he learned how to use electron diffraction (similar to X-ray diffraction, but using a beam of electrons). The paper was one of more than fifty that Pauling had published, yet he was still only 30 years old. In recognition of his astonishing energy and brilliance, he was awarded the Langmuir Prize as the most promising young scientist in the United States.

Pauling's book summarizing his work on bonds, published in 1939, became the most influential and widely read chemistry book of the century. By bringing quantum mechanics into his explanation of bonding, he explained how and why elements form compounds, and materials behave as they do.

Pauling used X-ray and electron diffraction, looked at magnetic effects, and measured the heat of chemical reactions, to calculate the distances and angles between atoms forming bonds.

He described hybridization, showing that the labels 'ionic' and 'covalent' are little more than a convenience to group bonds which really lie on a continuous spectrum from wholly ionic to wholly covalent.

He introduced the concept of electronegativity as a measure of the attraction an atom has for the electrons involved in bonding, and developed a table of values for different atoms. The electronegativity scale lets us say how covalent or ionic a bond is.

Finally, Pauling examined the way carbon forms bonds. As carbon has four filled and four unfilled electron shells it can form bonds in many different ways, making possible the

Chemical bonds

In a covalent bond, one or more electrons are shared between 2 atoms. So 2 hydrogen atoms form the hydrogen molecule, H2, by each sharing their single electron. The 2 atoms are bound together by the shared electrons. This was proposed by Gilbert Lewis and Irving Langmuir in 1916.

In an ionic bond, one atom gives away one or more electrons to another atom. So in common salt, sodium chloride, sodium gives away its 'spare' electron to chlorine. As the electron is not shared, the sodium and chlorine atoms are not bound together in a molecule. However, by losing an electron, sodium acquires a positive charge and chlorine, by gaining an electron, acquires a negative charge. The resulting sodium and chlorine ions are held in a crystalline structure. Until Pauling's explanation it was thought that they were held in place only by electrical charges, the negative and positive ions being drawn to each other.

myriad organic compounds found in plants and animals. The concept of hybridization proved useful in explaining the way carbon bonds often fall between recognized states.

The building blocks of life

Pauling's work on the carbon atom opened the door to the realm of organic chemistry. Organic compounds are the carbon-based chemicals on which all life forms on Earth are based. Inspired by X-ray photographs of organic compounds produced by Alfred Astbuy in England, Pauling turned his attention in the 1930s to the structures of complex organic molecules, an area previously of no interest to him.

X-ray diffraction alone is not a very useful tool for determining the structure of complex organic molecules, but it can show the general shape of the molecule. Pauling and his colleagues experimented with models to match possible chemical structures to the shapes revealed by X-ray diffraction. Sometimes they moved around simple flat shapes on paper; often, they needed more complex three-dimensional models.

The work was productive. Pauling demonstrated that the structure of haemoglobin (which carries oxygen in the blood) changes when oxygen is attached to it. He was one of the first people to explain how both antibodies and enzymes work. His work showed that physical chemistry, at the molecular level, could be used to solve problems in biology and medicine.

But the main challenge for biochemists was to understand proteins – the chemicals that control all processes in cells. Again, Pauling's work was key to progress.

The main challenge for biochemists was to understand proteins – Pauling's work was key to progress in this area.

The bond that makes life possible

The hydrogen bond is a special type of bond formed between a hydrogen atom and a nearby atom with a negative charge (see panel). Pauling did not discover it – the strange behaviour of some atoms around hydrogen had been observed before – but he did explain it in terms of quantum behaviour and he worked out a value for the energy involved. Once explained, it became clear that life on Earth depends on the existence of this small, weak bond, and many pieces of the biochemical jigsaw could at last fall into place.

Round and round

Pauling worked with Alfred Mirsky over an extended period, interrupted by the Second World War, to discover the structure of proteins. William Lawrence Bragg's team at the Cavendish Laboratory at Cambridge University, UK, was working on the same problem. This time it looked as though Bragg would get there first, but when he published his findings in 1950 it was clear that they were not quite right. Pauling came up with the correct answer, published the following year – the characteristic shape of a protein is a long chain twisted into a helix, or spiral, now known as the alpha-helix. The structure is held in shape by hydrogen bonds. He also explained the beta-sheet, a pleated sheet arrangement given strength by a line of hydrogen bonds. In May 1951, Pauling's team stunned the biochemical world by publishing the structure of seven fibrous proteins, including hair, silk and muscle.

Pipped at the post

The biggest, most complex and important biochemical molecule is DNA, the substance from which all chromosomes are made. In its structure, it encodes all the characteristics of a living organism. Tackling the structure of DNA was an obvious next step for both Pauling and Bragg, but neither of them would win this race. Instead the prize went to the British team of Crick and Watson (see p. 149).

Pacifist and activist

Pauling, along with many other leading scientists, had been asked to join the Manhattan Project to develop atomic weapons during the Second World War. He had once been a close friend of the leader of the project, Robert Oppenheimer, though their friendship had come to an abrupt end after Oppenheimer tried to persuade Pauling's wife to run away to Mexico with him. But it was Pauling's pacificism that made it impossible for him to take part, and which led to his difficulties with the US authorities in the years after the war as he campaigned relentlessly for control of nuclear arms. His passport was confiscated in 1952 and returned just in time for him to travel to Stockholm to collect his Nobel Prize for chemistry in 1954. In fact, the loss of his passport prevented him visiting England and seeing Rosalind Franklin's X-ray photographs of DNA. These might have led him, rather than Crick and Watson, to reveal the correct structure of DNA – a double helix, rather than the triple helix which Pauling suggested.

Pauling campaigned tirelessly to ban above-ground nuclear testing and to control nuclear weapons around the world. In 1958 he and his wife collected and presented to the United Nations a petition calling for an end to weapons testing that was signed by more than 11,000 scientists. Public pressure after this led to the Partial Test Ban Treaty signed in 1963 by the United States and Russia.

On the day the treaty was signed, Pauling was awarded the Nobel Peace Prize. This did nothing to help his position at home. His department at Caltech, still wary of his involvement, did not congratulate him, and only the biology department held a small party for him. The Senate Internal Security Subcommittee called him 'the number one scientific name in virtually every major activity of the Communist peace offensive in this country' and *Life* magazine called his prize 'an extraordinary insult to America'.

In later life, Pauling turned his attention to alternative medicine, in particular promoting large doses of vitamin C as a way of avoiding cancer. Pauling died, ironically from prostate cancer, on 19 August 1994 at his ranch near Big Sur, California. He was 93.

The hydrogen bond

When a hydrogen atom forms a bond with an atom which strongly attracts its single electron, little negative charge is left on the opposite side of the hydrogen atom. As there are no other electrons orbiting the hydrogen atom, the other side of the atom has a noticeable positive charge, from the proton in the nucleus. This attracts nearby atoms with a negative charge. The attraction – called the hydrogen bond – is about a tenth of the strength of a covalent bond.

In water, attraction between the hydrogen atoms in one water molecule and the oxygen atom in other water molecules makes water molecules 'sticky'. It gives ice a regular crystalline structure it would not otherwise have. It makes water liquid at room temperature, when other compounds with similarly small molecules are gases at room temperature.

Pauling took from quantum mechanics the idea of an electron having both wave-like and particle-like properties and applied it to hydrogen bonds. Instead of there being just an electrical attraction between water molecules, Pauling suggested that wave properties of the particles involved in hydrogen bonding and those involved in covalent bonding overlap. This gives the hydrogen bonds some properties of covalent bonds. He was finally proven correct by experimental evidence in 1999.

The DNA Team

Francis Crick, James Watson and Rosalind Franklin

DNA is probably the most famous molecule in the world – its double-helix shape the single most familiar image from biology and chemistry. It was discovered by Francis Crick, James Watson, and, more controversially, by Rosalind Franklin.

Francis Crick
1916–2004
James Watson
b. 1928
Rosalind Franklin
1920–1958

THE DISCOVERY OF THE STRUCTURE OF DNA, and the possibilities it has opened up, have brought complex science to public consciousness as much as any other scientific advance of the twentieth century. Understanding how DNA works has helped us to unravel the mechanism of inheritance, explaining how organisms inherit characteristics from their parents and ancestors. It has made possible new medical treatments, the controversial techniques of genetic engineering, cloning, screening for genetic disease and DNA fingerprinting.

Setting the scene

The understanding of inheritance began with the work of Gregor Mendel in the 1850s, who saw a pattern in how characteristics are passed on through generations of pea plants (see p. 99). Later in the nineteenth century other scientists saw chromosomes and discovered DNA in the cells, but it was not until 1944 that Oswald Avery realized that DNA carried the information that controls inheritance.

The work of Linus Pauling, William Bragg and others on the structure of proteins in the 1940s set the stage for exploring the structure of DNA. Pauling had revealed that the molecular structure of biological molecules accounts for their chemical behaviour. It was clear that understanding how DNA enables plants and animals to pass on genetic traits could only be based on a knowledge of its molecular structure. The race to discover the structure was on.

Open competition

In the United States, Pauling led the most experienced and best-financed research team. In the UK, the effort was split between the laboratory of King's College in London and the Cavendish Laboratory in Cambridge. The two British groups had different approaches. In London, Rosalind Franklin and Maurice Wilkins were working from X-ray images of DNA molecules, trying to work out the structure mathematically. In Cambridge, Francis Crick and James Watson made models of possible structures that would accommodate what was known about the chemical composition of DNA and the structures and sub-units already identified in other large molecules. The breakthrough came only when the fruits of both approaches were combined.

Cambridge ...

James Dewey Watson was born in Chicago on 6 April 1928. He was always an inquisitive child, and would not be silenced by a simple answer to a question. He spent a lot of time bird-watching with his father, and his first ambition was to be an ornithologist. He started studying for a degree in zoology at the University of Chicago at only 15 years old.

It was while studying for his PhD at Indiana University that Watson became interested in genetics. After his doctorate, he moved in 1950 to Copenhagen to study the effect of DNA on viruses. He become interested in the way X-ray crystallography was being used to reveal the shapes of complex molecules. As the best X-ray crystallography work was being done in England at the time, he set off to learn more, ending up in the Cavendish Laboratory in Cambridge.

Looking at lines

The technique of X-ray crystallography had been developed in 1912 by William and Lawrence Bragg, a father-and-son team working in England. The Braggs discovered that if X-rays were directed at a crystalline structure, a regular pattern of lines was produced. X-rays are scattered by striking atoms in the crystal, or pass straight through when there are no atoms in their path. Using X-rays of known wavelengths, diffraction patterns – the pattern of lines – could be used to work out the arrangement of atoms in a crystal lattice.

X-ray crystallography is the interpretation of X-ray diffraction patterns in order to discover the structure of crystals and molecules. It was first applied to biological molecules by Linus Pauling.

Watson's colleague in Cambridge, Francis Harry Compton Crick, came to biology relatively late in life. Born in the north of England on 8 June 1916, he was an inquisitive child with a passionate interest in science. He read all the science books he could get hold of and experimented at home in his kitchen. He studied physics at University College in London, then worked for the Admiralty on mines during the Second World War. After the war, he looked around for something more interesting to do. He went first to the Strangeways Laboratory in Cambridge, and soon after moved to the Cavendish to study proteins. When Watson arrived in Cambridge, the two immediately became firm friends. They shared an office together and although both were supposed to be concentrating on other projects, their shared interest in DNA led them to work together on modelling its structure.

Watson: 'The instant I saw the picture [of the DNA molecule] my mouth fell open and my pulse began to race.'

... *versus* London

To call the pair working in London a team implies a degree of co-operation that did not exist. Maurice Hugh Frederick Wilkins was born in Pongaroa, New Zealand, in 1916. He moved to England at the age of six, and later studied physics at Cambridge University. During the Second World War he work for a time on the Manhattan Project in the United States. After the war he returned to Britain. He worked first as a physics lecturer at St Andrew's University in Fife, Scotland, and then moved to King's College to work on biological molecules, including DNA and viruses.

Rosalind Elsie Franklin was born in London, England, on 25 July 1920. She was

The discoverers of the structure of DNA. James Watson (left) and Francis Crick with their model of part of a DNA molecule in 1953.

a brilliant child, and her prosperous background meant that she was lucky enough to go to one of the few schools that taught physics and chemistry to girls at the time. Her father discouraged her ambitions to become a scientist; even so, she went to Cambridge University to read chemistry in 1938. After finishing her degree, she spent a year in research at Cambridge, but gave it up to work in industry studying the physical structure of coal. In 1947 she moved to Paris where she learned about X-ray crystallography. She returned to England in 1951 and was given a job at Wilkins's laboratory in King's College. Unfortunately, she was employed while Wilkins was away and they got off to a bad start when he returned and assumed she was an assistant rather than his peer, appointed to work on the same problem as he was already exploring. Their relationship only deteriorated.

Working methods

Franklin's X-ray photographs were brilliant – the best produced by anyone. She enjoyed the additional good luck of having access to the best sample of DNA in the world. This had been derived from the thymus glands of calves by a Swiss scientist and generously given to Wilkins. Now, undiplomatically, it was handed over to Franklin by the director of the laboratory, John Randall.

X-ray crystallography can show the general shape of a molecule, but not the type of each atom or molecule in it. Crick and Watson tried to make physical models of the structure of DNA, first by using cardboard cut-outs and then metal plates, screws, rods and coloured balls to represent groups of atoms and the bonds between them. Their aim was to piece together the possible arrangements of the atoms from their knowledge of the chemical composition of DNA, choosing structures that would match the evidence from X-ray photographs.

In the United States, Pauling was doing much the same thing. His unrivalled knowledge of chemical bonds should have given him the edge, but he was hampered by having only poor-quality X-ray images to work from.

The idea that the structure was some form of helix was already gaining ground.

Pauling had already discovered the alpha-helix structure in proteins, and the early photographs could be interpreted as supporting the theory. Franklin, however, dismissed the idea that a helix was the key to the structure along with the practice of making models, which she derided, preferring her own methods of experimentation and measurement.

Stumbling progress

Watson first saw Franklin in 1951 when she was presenting some of her X-ray photographs and showing the basic size and shape of the DNA strands at a meeting in London. He did not pay enough attention to be able to use the information properly, though. Crick and Watson built a model based on Watson's imperfect recollection of Franklin's evidence which showed the molecule as a triple helix. It was badly flawed, which Franklin gleefully pointed out to them when they demonstrated their model. A better knowledge of chemistry might have saved them from the humiliating mistake. Crick and Watson suggest they join forces with Franklin, but she refused. The director of the Cavendish, embarrassed and exasperated, told the pair to stop working on DNA and leave the problem to the London laboratory. They pretended to comply.

Franklin continued working on her own idea of the structure, largely alone, as she would not co-operate with Wilkins. She had identified two forms of DNA, varying with humidity. She set aside her most stunning photograph of the wet form, which she called Photo 51, as she was more interested in the dry form. She worked slowly, determined to work out the solution from photographs and calculations alone.

Linus Pauling, working in the United States without the benefit of Franklin's X-ray photographs, came up with a triple helix structure in January 1953. He wrote to his son, working at the Cavendish, with the news and sent him a draft of his paper explaining the structure. Crick and Watson, at first distraught at being beaten to the answer, were relieved to see that Pauling had got it wrong, making the same mistake they had made earlier.

Breakthrough

Before leaving King's College to work at Birkbeck on viruses, Franklin gave a final seminar in which she stressed her belief that the structure of DNA was not a helix. Watson visited her a few days later to show her Pauling's paper but the two quarrelled. Afraid she was going to hit him, Watson quickly left. He bumped into Wilkins, who was already frustrated by the delay. Wilkins showed him Franklin's best photograph, probably Photo 51, without her knowledge. It clearly showed that the structure must be a helix. 'The instant I saw the picture my mouth fell open and my pulse began to race,' Watson remembers. He sketched what he had seen in the margin of a newspaper on the train back to Cambridge. It took only a month for Crick and Watson to perfect their model after that.

The famous double helix

Crick and Watson's final model shows a double helix made of two joined chains of nucleotides. Two outside strands are held at the same distance apart by pairs of nucleotide bases interlocking in the centre, acting like the rungs of a ladder. The bases are always paired in the same way: adenine with thymine, and guanine with cytosine.

Revealing the structure of DNA did indeed show how it works. The two strands separate, then each acts as a template, allowing the other to regenerate – where an adenine appears, a thymine must be added opposite it, and so on. As cells divide to multiply, the DNA is copied exactly into each new cell.

Moving on

Crick was awarded his PhD from Cambridge in 1953 and continued to work on the mechanism of DNA with Watson until 1966, when he turned to embryology. Later he followed up an early interest in how the brain works and turned his attention to consciousness and neural networks. In later life, he came to champion some less mainstream, often unpopular ideas. These included the theory that life on earth originated in outer space and was sent here by intelligent beings in unmanned rockets, and his belief that the knowledge of genetics should be used to perfect the human race. He died on 28 July 2004.

Watson continued to work in genetics, teaching in the United States at Harvard and Caltech. He became head of the Human Genome project in 1988.

Wilkins continued to teach in London, and to campaign against nuclear weapons. He died on 5 October 2004.

Despite never receiving credit for her role in solving the mystery of DNA, Franklin was pleased that the structure had been discovered. She worked on viruses at Birkbeck College until her early death from cancer on 16 April 1958.

Nobel Prize – but not for all

The structure of DNA as demonstrated by Crick and Watson was accepted immediately. But by the time a Nobel Prize was awarded for the work, Rosalind Franklin had died. Nobel Prizes are not given posthumously; the portion of the prize which might otherwise have been hers was awarded to Maurice Wilkins, along with Francis Crick and James Watson, in 1962.

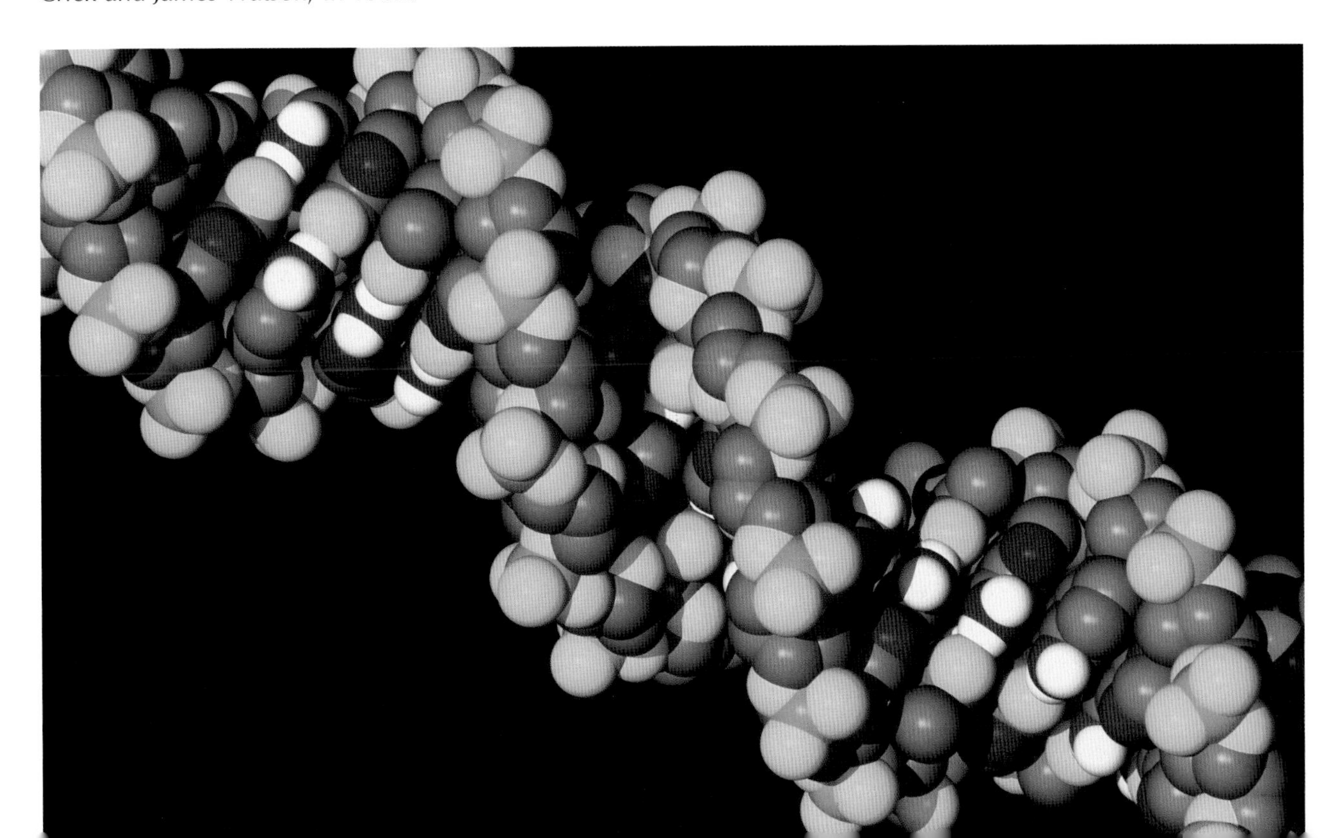

The structure of a DNA molecule – the famous double helix.

Stephen Hawking

The English cosmologist Stephen Hawking (1942–) is one of the leading scientists alive today, and his brilliant work on black holes in space and the big bang has profoundly influenced our understanding of the universe.

b. 1942

A CENTURY AGO, ASTRONOMERS BELIEVED THE UNIVERSE was little bigger than our own Milky Way galaxy of stars, and was completely stable and unchanging. But over the first few decades of the twentieth century, that view was profoundly shaken.

First of all, powerful telescopes revealed that there are countless galaxies beyond the Milky Way; then, in the 1920s, the great American astronomer Sir Edwin Hubble discovered that all these galaxies are speeding away from us, which means that, far from being stable, the universe must be expanding at a phenomenal rate.

Meanwhile, two great concepts – quantum physics and Einstein's theory of relativity – turned classical physics, the seventeenth-century physics of Newton, on its head. These extraordinary ideas only gradually made a general impact on the world of science, but a number of scientists immediately saw some implications. In 1917, for instance, ten years before Hubble observed the expansion of the universe, the Russian astronomer Aleksandr Friedmann inferred this expansion from Einstein's relativity. Einstein actually disagreed with Friedmann, and was shocked when Hubble's discovery proved Friedmann right.

As it happens, this was already the second time Einstein's own interpretation of relativity had been challenged, again rightly as it turned out. A year earlier, in 1916, the German astronomer Karl Schwarzschild used Einstein's theory to work out what happens when a star collapses under the force of its own gravity. Schwarzschild concluded that, as the star contracts, its gravity grows so powerful that nothing, not even light, can escape. It becomes a 'black hole' in space. Such a hole was later found to be centred on a minute point called a singularity, where time and all forces become one. The size to which a star must shrink before it becomes a black hole is called, appropriately, the Schwarzschild radius and is about 3 km for a star the size of our Sun.

Over the next half century, scientists began mathematically to wind back the clock of the expanding universe, and they realized that, although it is now big, it once must have been very very small. The theory was that it burst into existence and swelled outwards about 13 billion years ago in what came to be called the big bang. The big bang theory soon became firmly established, even though understanding of the processes involved was shaky.

Black holes, however, remained controversial. After all, they could not, by definition, be seen. Some Soviet scientists argued that they could not exist at all because they depended on the star collapsing perfectly symmetrically, which, they argued, was highly unlikely.

Significantly, while Einstein's relativity had played a key part in both the big bang and Black Hole theories, the other great revolutionary idea, quantum physics, seemed almost to have been sidelined as irrelevant to cosmology. This was because it apparently refers only to the minute, subatomic level, and not to the scale of the universe.

It was Stephen Hawking's brilliant insight that brought the big bang and black holes – relativity and quantum physics – all together to give an extraordinary theoretical picture of cosmic forces at work.

First of all, as a young graduate, Hawking realized that the big bang might be a black hole in reverse, expanding from a singularity. This gave cosmologists the mathematical tools to develop a fuller picture of the origins of the universe.

Then, in the early 1970s, Hawking realized that quantum effects might apply to the 'event horizon' or rim of black holes (see box on p. 157). If they did, he argued, they would make a black hole glow faintly – and so perhaps be detectable after all, making this hitherto theoretical idea a reality. This glow came to be called Hawking radiation. Even more importantly, by bringing quantum physics into the study of black holes, Hawking had drawn it into the whole cosmological field and so opened the way to an all-embracing physical theory of the universe. It is this that Hawking and his colleagues are looking for at the moment.

Bright little boffin

Stephen Hawking is now famous not just for his cosmological theories, but the terrible disease which has left him completely paralyzed and only able to speak with a synthesized voice. The degenerative disease that paralyzed him is called amyotrophic lateral sclerosis (ALS), which damages the nerve cells in the spinal cord that control the body's voluntary muscles. When diagnosed with the disease in the 1960s at the age of 22, Hawking was given just a few years to live. Although the disease has progressed, he has defied all expectations and remained alive – and as clear in his mind as ever.

When he was a child, there was little sign of the disease. He was born on 8 January 1942 in Oxford, where his parents had moved temporarily to escape the London Blitz. His date of birth was, coincidentally, the 300th anniversary of Galileo's death, and close to the 300th anniversary of Newton's birth. When Stephen was 8, the family moved to St Albans, just outside London. By all accounts, he was quite a withdrawn child. Although his true brilliance was not at first apparent, a friend recalls how he began to develop from 'a bright little boffin' to someone who had 'an overarching arrogance, if you like, some overarching sense of what the world was about'.

It was Stephen Hawking who brought the big bang and black holes, relativity and quantum physics all together.

At the young age of 17, he took the entrance exam for Oxford University and was awarded a scholarship to study natural science. Emerging with a first-class degree three years later, he went to Cambridge to study cosmology with, he had hoped, the eminent Fred Hoyle. Unfortunately, Hoyle was too busy to give him any time, and Hawking's pride took a knock. His sense of self took an even more devastating blow when, a few months after arriving in Cambridge, after suffering a few falls, he was diagnosed with ALS.

Marriage and progress

It was about this time, though, that he met and fell in love with 18-year-old Jane Wilde. It may have been her influence that gave him a new sense of purpose, but he became determined to accomplish things despite his illness. 'I dreamt that I was going to be executed,' he remembers. 'I suddenly realized there were a lot of worthwhile things I could do, if I were reprieved.' Two years later, he and Jane married, and she remained his support for 25 years. Against all the odds, they had three children together.

In the meantime, Hawking was beginning to make a name for himself. First of all there was the time when Fred Hoyle was giving a lecture at the Royal Society on his steady-state theory, insisting the universe is not forever expanding but steadily hovering between expansion and contraction. At the end of the lecture, after the cheers for Hoyle had died down, the young graduate Hawking stood up and said, 'The quantity you are talking about diverges.' If it was so, Hoyle's argument that day

An energy-releasing black hole. The energy around this black hole was measured by astronomers and found to be greater than expected. The extra energy is thought to be rotational energy extracted when the black hole's spin is slowed by the twisting of magnetic field lines close to its border (event horizon).

Black holes and uncertainty

Hawking, more than anyone, has demonstrated the power of mathematics in working out the cosmos. While thinking about black holes back in the 1970s, Hawking began to notice a strange similarity between the event horizon of a black hole and the second law of thermodynamics. This is the famous law that says that an isolated system will always tend to gain entropy (become more chaotic), and it will never become more ordered if left to itself. Hawking said it was like a house: if it stops being repaired, it gradually falls apart. In the same way, Hawking realized the surface of a black hole can only stay the same size or swell – it can never shrink or become more ordered.

To understand why this might happen, Hawking took into account Heisenberg's Uncertainty Principle, formulated in 1927. Heisenberg showed that it is not possible to be certain of both an object's momentum and its position at the same time. This is because the way of determining the question will distort either one or the other. On an everyday scale, this distortion is so small that it does not matter. But on the level of subatomic particles it is crucial, and leads to all kinds of weird 'quantum effects' in which particles apparently ignore the rules of classical physics and jump and pop up all over the place, like rabbits from a magician's hat. Just how this can happen is difficult to explain, but quantum effects have proved to be real and are the basis of technology such as lasers.

One of the most astonishing aspects of Heisenberg's Uncertainty Principle is what it says about empty space. In fact, it says there is no such thing as empty space, because empty space would be a precise state and there is no such thing as a precise state. To create what is probably empty space, pairs of 'virtual' particles must oscillate either side of the zero that is empty space. These pairs are positive particles and negative antiparticles. When they come together, they annihilate each other, but they are constantly bobbing in and out of space.

Hawking realized that this bobbing and popping of virtual particles was going on all along a black hole's event horizon, which is the edge of space. Negative particles would be drawn into the black hole and positive particles pushed out. The negative particles are what stop the size of the black hole ever decreasing. The expelled positive particles emerge as heat, miniscule – just a few millionths of a degree above absolute zero – but in theory measurable. So black holes are not black, but emit heat radiation, which came to be called Hawking radiation.

Hawking went further and suggested that just as a star losing radiation gradually dwindles, so a black hole would eventually evaporate into pure radiation – that is, it would explode. His ideas were published in the magazine *Nature* in a paper entitled 'Black Hole Explosions?', which is now acknowledged as one of the classics of cosmology.

was useless. 'Of course, it doesn't diverge,' Hoyle replied. 'It does,' Hawking insisted. 'How do you know?' 'Because I worked it out.' And he had.

Delving into black holes

Hawking then began to study the young Oxford mathematician Roger Penrose's work on black holes. Penrose had shown that right at the centre of a black hole, inside the event horizon, there must be a point where all the mass shrinks to nothing – a point called a singularity. Hawking turned this idea round to look at the origins of the universe and suggested that the big bang was basically a black hole in reverse, and that it must have begun as a singularity, an infinitesimally small point containing the entire substance of the universe.

The significance of Hawking's ideas was immediately appreciated among the cosmological community, and his reputation grew even as his body began to deteriorate. His speech was beginning to slur, and he became barely able to write. His wife Jane bore the brunt of his frustration, took dictation and typed his scrawled notes. By 1974, the year he was elected to the Royal Society as one of its youngest fellows ever, he was able to move only in a wheelchair, and his voice had dwindled to a moan that was intelligible only to his family, friends and close colleagues.

Yet if his body was weakening, his mind was not. In the late 1970s, he made what is perhaps his most famous discovery, showing not only that black holes might be detectable, but that they might eventually explode (see box on p. 157) This was such a radical idea that not everyone accepted it at first, and a few cosmologists still challenge it.

A Brief History of Time

In the early 1980s, Hawking began to dictate ideas for a popular book on cosmology, partly to earn money to pay for his children's schooling. He finished it in 1985, and went to Geneva to visit the CERN particle accelerator while Jane took a much-needed vacation, leaving his nurse and a research assistant to look after him. A few days later Hawking was found struggling to breathe and rushed to hospital. His windpipe had been blocked by pneumonia, and the only way to save his life was to give him a tracheotomy, which meant he would not be able to speak again. Jane had hurried back to be at his bedside.

The Hawkings returned to Cambridge, with Stephen only able to communicate by blinking his eyes. As news of his plight spread, Californian computer expert Wal Woltosz offered the help of a computer package that would synthesize a voice with just a tiny finger movement. It required long practice, but eventually Hawking mastered it, and acquired the synthesized voice that is familiar to many people today.

Hawking's book, *A Brief History of Time,* was published on April Fool's Day 1987. To everyone's astonishment, it was a great success, and quickly became the best-selling science book in history, despite its complex subject matter. Nobody quite knows why it did so well, but perhaps many people felt that this strange, brilliant man might just reveal some of the ultimate truths about our universe. It seemed important to be in the know. In the final chapter, Hawking talks about the nature of God.

Recent years

Hawking was now a major celebrity, and as he continued to think about theories that would bring all the laws of physics together in one simple equation, he also began to tour the world with his nurse Elaine and make countless TV appearances as the ultimate science pundit. A TV film was made about him called *Master of the Universe*. The strain on his marriage with Jane began to tell. In 1990, the couple broke up and Hawking moved in with his nurse Elaine.

Throughout the 1990s, Hawking continued to work on his 'Theories of Everything', and some people wondered if his greatest achievements were behind him. Then in July 2004, he made a stunning announcement to a Dublin conference.

For decades he had debated with other scientists over the 'information paradox', which, following on from his quantum view of black holes, was the question of whether or not data might be recoverable from black holes. Hawking was so sure that it could not that he had a bet with Kip Thorne of Caltech, who thought it could. The conference was stunned when Hawking announced he had just solved the paradox – and he had lost his bet. It seems that quantum fluctuations, like those that produce Hawking radiation, will in fact allow data to leak out to the outside. In other words we can, in theory, discover exactly what goes on inside a black hole.

Index

Picture Credits

AKG IMAGES – 16, 18, 37, 74, 95, 103, 115, /ERICH LESSING 60
BRIDGEMAN ART LIBRARY – 27
CORBIS – 10, 15, 17, 20, 26, 35, 43, 44, 50, 58, 68, 72, 84, 86, 93, 119, 141, 144, 147, 154
HULTON – 25, 45, 57, 117
MARY EVANS PICTURE LIBRARY – 54
SCIENCE PHOTO LIBRARY – 31, 41, 59, 77, 80, 91, 94, 107, 111, 124, 128, / A BARRINGTON BROWN: 151, /ADAM HART-DAVIS: 88, /AMERICAN INSTITUT OF PHYSICS: 140, /BERNHARD EDMAIER: 69, /CERN: 125, /CHRISTIAN DARKIN: 53, /DAVID A. HARDY: 138, /DR JEREMY BURGESS: 23, 55, 89, /ERIC HELLER: 113, /FRANCES EVELEGH: 62, /JAMES KING-HOLMES: 99, /JEAN-LOUP CHARME: 22 , /PROF. K. SEDDON & DR. T. EVANS, QUEENS UNIVERSITY BELFAST: 153, /MEHUA KULYK: 33, 105, /NASA: 156, /NIELS BOHR ARCHIVE/AMERICAN INSTITUTE OF PHYSICS/M. BOHR COLLECTION: 132, /NRAO/AUI/NSF: 120, /PETER MENZEL: 81, /PHILIPPE GONTIER/EURELIOS: 97, /SANFORD ROTH: 136, /SCOTT CAMAZINE: 133, /SEYMOUR: 109, /SHEILA TERRY: 39, 49, 64, 76, 101, /SIMON FRASER: 129, /SINCLAIR STAMMERS: 46, /STEVE ALLEN: 30, /TONY CRADDOCK: 123, /VOLKER STEGER: 92
TOPHAM PICTUREPOINT – 66

THE GREAT SCIENTISTS

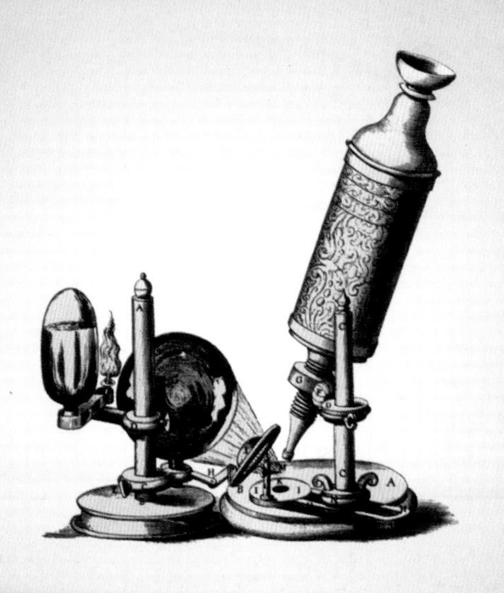